Nuclear Physics: Energy and Matter

NUCLEAR PHYSICS: ENERGY AND MATTER

J M Pearson

Université de Montréal

ADAM HILGER LTD, BRISTOL AND BOSTON

Samford University Library

© Adam Hilger Ltd 1986

All rights reserved. No part of this publication may be reproduced, stored in a retrieval system or transmitted in any form or by any means, electronic, mechanical, photocopying, recording or otherwise, without the prior permission of the publisher.

British Library Cataloguing in Publication Data

Pearson, J. M.
Nuclear physics: energy and matter.
1. Nuclear physics
539.7 QC776

ISBN 0-85274-804-3

Consultant Editor: **Dr W D Hamilton**, University of Sussex

Published by Adam Hilger Ltd
Techno House, Redcliffe Way, Bristol BS1 6NX, England
PO Box 230, Accord, MA 02018, USA

Typeset by Mathematical Composition Setters Ltd, Salisbury, UK
Printed in Great Britain by J W Arrowsmith Ltd, Bristol

QC
776
.P34
1986

I dedicate this book to the memory of my parents,
KATHLEEN MARY DAWSON and CHARLES VERNON PEARSON

CONTENTS

PREFACE

There exist already many introductory nuclear physics textbooks, and the appearance of yet another requires some justification. The *raison d'être* of the present book lies in the author's discontent with the objectives of the traditional first course in nuclear physics, for which most of the existing texts cater. The tendency seems to be for such courses to be directed primarily towards students who intend to specialise in this field later on, overlooking the fact that for the vast majority of students at this level the introductory course will also be the last one.

Actually, these two different potential audiences will be discernible in all courses in specialised topics at this level, e.g. particle physics, solid state, plasma, etc, but in most cases there may be no great conflict between the needs of the two groups of students. However, one has the impression that in the case of nuclear physics there is an acute problem, arising from the fact that the greatest scientific significance of nuclear physics lies in its extrinsic aspects, i.e. in its relationship with other fields, rather than in its intrinsic aspects. I am referring here in the first place to the large-scale, macroscopic manifestations of nuclear physics, since these, being associated with the generation of energy and the creation of matter, have a profound influence both on the face of the natural world, and, increasingly, on the man-made world. The energy of the stars, including that of the sun, is nuclear in origin, and in the course of this energy being generated within a star nuclei of ever greater complexity are synthesised. It is in this way that nearly all the chemical elements heavier than hydrogen have been formed; the exceptions are some very light nuclides synthesised in the primordial 'big bang'. Thus the very existence of matter as we know it depends on the laws of nuclear physics, and the way in which they operate in the stars, and in the big bang.

If, for example, the basic interaction between nucleons were only slightly different it would be impossible to form a stable nucleus containing six protons, in which case there would be no carbon in the universe, and hence no life either, at least not in any form in which we know it. And in fact, if carbon did not exist it is most unlikely that any of the heavier elements would be found either, since their synthesis in the stars depends on the presence of carbon. Thus we see that both chemistry and biology depend ultimately on nuclear physics.

As for the artificial large-scale manifestations of nuclear physics that have been alluded to, I have in mind, of course, the fission chain reaction†, and man-made thermonuclear fusion reactions, which are essentially of the same kind as those taking place in the stars. Since both these self-sustaining processes constitute sources of enormous amounts of energy, they offer the brightest hopes, and at the same time pose the most appalling threat to mankind. (It is sometimes asserted that these two processes constitute the only sources of terrestrially available energy that is not solar in origin. However, geothermal power is partially non-solar, since radioactive decay contributes to the heating of the earth's interior. In any event, it is indisputable that *all* our energy is basically nuclear. At the same time it must be remembered that the radioactive elements whose decay heats the earth, and the uranium that powers our reactors, were synthesised in the stars. Thus we can say that all our energy is ultimately both nuclear and *stellar*, if not solar.)

These large-scale manifestations constitute, then, the first major interface of nuclear physics with other domains of science. We have spoken of the crucial role played by the basic force between nucleons in determining firstly the properties of nuclei and through them the large-scale manifestations. Instead then of looking outwards from the nucleus through a telescope ('macroscope'?) let us rather look inwards through a microscope, and enquire as to the origin of the force between nucleons. It is at this point that we perceive the second major interface of nuclear physics with another branch of physics: that with elementary particle, or high-energy physics ‡. Clearly, nuclear physics itself is a manifestation of the fundamental laws of physics.

Actually, although some allusion to this latter aspect of nuclear physics is made in Chapter I, we do not dwell very much on it in this book: it belongs properly to a specialised course in elementary particle physics, and we shall be concerned overwhelmingly with the first interface, i.e. the large-scale manifestations. Nevertheless, the reader should bear in mind the existence of both these interfaces, and realise the extraordinary pivotal role played by nuclear physics: it links the world of ordinary matter in which we live with the transcendental world of elementary particle physics, and thus with the fundamental laws of physics.

†This was originally discovered as a man-made phenomenon, but recently it has been found to occur naturally in the earth (see § 24).

‡The terms 'elementary particle physics' and 'high-energy physics' are used synonymously in the sense that most elementary particles are highly unstable, and so have to be created if they are to be studied in a free state. This, of course, requires very high energies, either in the form of accelerators or cosmic rays.

The profound importance of the large-scale manifestations of nuclear physics comes to be appreciated by most nuclear physicists sooner or later in their careers, and many find in it the main intellectual motivation for their activity†. However, this appreciation will not be acquired in the traditional introductory course, and the result is that the vast majority of physics undergraduates will *never* acquire it. Not the least of the several unfortunate aspects of this situation is the spectacle of large numbers of physics graduates leaving the universities and going out into the wide world unable to make any scientific contribution to the great public debate on the future of nuclear energy, a debate for which it is imperative that there be as much informed input as possible, so crucial is it to the future of mankind. (High-school physics teachers are particularly well placed to educate the public in these matters, and if they cannot do so this can only be blamed on their university training.) At the same time one must deplore the possibility of being able to obtain a university degree in physics without any awareness of how the laws of nuclear physics, as established in the laboratory, enable us to understand how the distant stars shine. This marriage of nuclear physics with astronomy must be regarded as one of the great intellectual achievements of our time: one of the newest of sciences is giving us a very deep insight into a mystery that has fascinated the mind of man from time immemorial.

In writing the present elementary nuclear physics text, then, particular emphasis has been placed on these various 'macroscopic' aspects of the subject. Nevertheless, I would insist that it is not to be regarded as an introductory text to nuclear astrophysics, and still less to reactor theory, even if these two topics do receive a much larger share of attention than is usual. Actually, it is only in the last two chapters of the book that these subjects are taken up in any detail, the first four chapters covering fairly conventional material, even if there is a tendency to orient this material towards the eventual 'macroscopic' applications. This latter tendency is particularly evident in the emphasis we place on the semi-empirical mass formula and the compound-nucleus reaction mechanism (in this connection we remark that the underlying liquid drop and shell models are discussed in some detail, as is the relationship between them).

At the same time some topics that appear in most introductory texts are not presented at all here. I am thinking particularly of nuclear magnetic moments and effective-range theory, topics that are absolutely essential for those who intend to specialise in nuclear physics, but irrelevant to anyone who wishes simply to acquire some idea of what nuclear physics is about and where its scientific significance lies. Indeed, most introductory nuclear physics courses probably devote an excessive amount of time to these

† Others regard the nucleus primarily as a laboratory for testing the properties of the fundamental interactions.

topics, with the result that many students are not only deterred from studying nuclear physics any further but actually develop a positive hostility towards it. Since the specialist has to rework such topics later on anyway, it would seem that they can be safely excluded from a first course and replaced by some of the less conventional material to be found in the present book.

A knowledge of the elements of quantum mechanics and a familiarity with the basic features of atomic structure are presupposed. It should be possible to cover most of the material within the span of a one-semester course, at the end of which it is to be hoped that the student will have become more aware of the general scientific relevance of nuclear physics than is customary. Even if this requires foregoing the detailed study of certain special topics, it should be well worthwhile.

Many people have helped me in the realisation of this book. I am particularly indebted to G C Hanna and M Arnould for their extensive and detailed criticisms of the original drafts of Chapters V and VI, respectively. B Rouben, L Amyot and G Beaudet also made valuable comments on these two chapters, while W Del Bianco, P Depommier, B Goulard, L Lessard, J LeTourneux, G Paquette, A Richter and P Taras scrutinised various other sections of the book at different stages of development. The entire manuscript was read by W D Hamilton, who made many helpful suggestions. I have also benefited considerably from the comments of several generations of students at the University of Montréal who were exposed to successive revisions of the text. To all these people, and to the several colleagues in various places who encouraged me in this project, I express my gratitude.

The bulk of the typing was done by Lucie Roux and Lise Blanchet, with Danielle Boileau-Charette and Pauline Vézina participating at various stages; I am grateful to all four for their skill and patience. Likewise I thank J Bérichon for his execution of the figures. Finally, I wish to express my thanks to my daughter, Susan A Pearson, for compiling the index and assisting with the proof reading.

J M Pearson

General Bibliography

The following titles refer, to a greater or lesser extent, to all chapters of this book. Specialised bibliographies are given at the ends of Chapters V and VI.

Bowler M G 1973 *Nuclear Physics* (Oxford: Pergamon)

Burcham W E 1973 *Nuclear Physics, an Introduction* 2nd edn (London: Longman)

Cohen B L 1971 *Concepts of Nuclear Physics* (New York: McGraw-Hill)

Elton L R B 1959 *Introductory Nuclear Theory* (London: Pitman)

Enge H A 1966 *Introduction to Nuclear Physics* (Reading, MA: Addison Wesley)

Frauenfelder H and Henley E M 1974 *Sub-Atomic Physics* (Englewood Cliffs, NJ: Prentice-Hall)

Glasstone S 1967 *Sourcebook on Atomic Energy* 3rd edn (New York: Van Nostrand)

Irvine J M 1975 *Heavy Nuclei, Superheavy Nuclei, and Neutron Stars* (Oxford: Oxford University Press)

Marmier P and Sheldon E 1969 *Physics of Nuclei and Particles* 2 volumes (New York: Academic)

Preston M A and Bhaduri R K 1975 *Structure of the Nucleus* (Reading, MA: Addison-Wesley)

Roy R R and Nigam B P 1967 *Nuclear Physics, Theory and Experiment* (New York: Wiley)

Segrè E 1977 *Nuclei and Particles* 2nd edn (New York: Benjamin)

Physical Constants

Avogadro's number $(N_A) = 6.022169 \times 10^{26}$ kmol^{-1}
Boltzmann's constant $(k) = 1.38062 \times 10^{-23}$ J K^{-1}
$(= 8.61708 \times 10^{-5}$ eV K^{-1})†
Compton wavelength of electron $(\lambda\!\!\!^{-}_e \equiv \hbar/m_e c) = 3.86159 \times 10^{-13}$ m
Electron charge $(e) = 1.60219 \times 10^{-19}$ C
$(e^2 = 1.4400$ MeV fm)†
Fine-structure constant $(\alpha \equiv e^2/\hbar c) = 1/137.036$
Gravitational constant $(G) = 6.6732 \times 10^{-11}$ m^3 kg^{-1} s^{-1}
Planck's constant/2π $(\hbar) = 1.054592 \times 10^{-34}$ J s
Rest mass of electron $(m_e) = 9.10956 \times 10^{-31}$ kg
Rest mass energy of electron $(m_e c^2) = 0.511004$ MeV†
Rest mass of neutron $(M_n) = 1838.64\ m_e$
Rest mass of proton $(M_p) = 1836.11\ m_e$
Velocity of light *in vacuo* $(c) = 2.99793 \times 10^8$ m s^{-1}

†In addition to SI units we have used:
1 fm (fermi ≡ femtometre) $= 10^{-15}$ m
1 eV (electron volt) $= 1.60219 \times 10^{-19}$ J

CHAPTER I

GENERAL INTRODUCTION TO THE NUCLEUS

§1 ORIGINS†

Although speculation on the atomic structure of matter goes back to Democritus and his school in the Greece of the fifth century BC, it was only with the work of Dalton at the beginning of the last century that the concept of atoms became scientific, in the sense that it could be used not only to correlate a large amount of data, but also to make quantitative predictions. It should be said, though, that Dalton benefited greatly from the considerable progress that had been made in the eighteenth century towards elucidating the concept of the chemical element, since he arrived at his conclusions essentially through measuring the relative masses of various elements that entered into chemical combination.

The concept of atoms formed in this way was of necessity somewhat abstract, and at first there was no indication at all as to their dimensions or structure. However, during the course of the nineteenth century, the physical reality of atoms became more and more apparent, mainly as a result of the successes of kinetic theory and statistical mechanics. In particular, by 1870 it had become possible to estimate the diameter of atoms: viscosity measurements in gases indicated a value of the order of 10^{-8} cm. Even so, opposition to the very notion of atoms continued until as recently

†For a fuller discussion see the book by Glasstone cited in the General Bibliography.

as 1900, especially from Ostwald, and it was only overcome once and for all by two separate discoveries which showed that atoms had structure, and could no longer be regarded as the 'hard, indivisible and immutable objects' that tradition demanded.

In the first of these discoveries, J J Thomson (1897) found that the charge/mass ratio e/m of the so-called cathode rays in discharge tubes was the same for all gases ($\sim 2 \times 10^{11}$ C kg^{-1}). He was thus led to identify the negatively charged *electron* as a universal constituent of all atoms†. Since atoms are normally electrically neutral it follows that there must be another component which is positively charged; clearly, in discharge-tube phenomena atoms are being broken up, or 'ionised', into these two components of opposite charge. The positively charged component in this discharge-tube ionisation was identified as the canal rays emerging from behind the cathode, and it was found that for a given elementary gas the charge/mass ratio could take different values, but was always an integral multiple of some unit. Furthermore, this unit of charge/mass ratio for the canal rays varied from one gas to another, and in any case was very much smaller than the corresponding ratio e/m for the electrons.

The picture of the atom that emerged from this was highly asymmetrical with respect to charge: while there could be several negatively charged electrons (any number of which could be removed in ionisation) there was but one positively charged component, and this carried nearly all the mass of the atom. J J Thomson imagined the atom to be more or less homogeneous, consisting of a uniformly and positively charged sphere of constant density, in which the electrons were embedded. This so-called 'plum-pudding' model of the atom was about the only one that was consistent both with classical physics and the facts as they were known before 1911. In particular, atomic line spectra could be understood quite simply in terms of the natural oscillations of the electrons that could be excited within such an atom.

The other key development to which we have alluded, leading to an even more radical revision of traditional ideas, was Becquerel's discovery in 1896 of *radioactivity*, the spontaneous emission by certain minerals of strongly ionising radiations. Three types of radioactivity were soon recognised: α (alpha) rays, identified by Rutherford (1909) as ionised helium atoms (actually ^{4}He, see § 3); β (beta) rays, identified by Becquerel (1900) as electrons and γ (gamma) rays, identified by Rutherford (1914) as electromagnetic radiation.

All three types of radioactivity were shown to be quite independent of temperature and pressure and were thus interpreted as atomic rather than

†Actually, the facts of electrolysis had already suggested the electrical nature of atoms, and indeed even the existence of a fundamental unit of electricity, to which the name 'electron' had been applied in 1891.

molecular processes, i.e. the emission of the ionising radiation involved a change in atomic structure. Furthermore, this change in atomic structure was shown by Rutherford and Soddy, working at McGill University, Montreal, at the turn of the century, to be far more radical than the changes accompanying line spectra. Extensive chemical studies of various radioactive minerals led inexorably to the startling conclusion that in emitting α and β radiation (but not γ), atoms of one chemical element actually transformed spontaneously into those of another, a notion which ran counter to all traditional views on the nature of elements and atoms.

These radiochemical studies also led to the discovery that some radioactive elements appeared to be chemically indistinguishable from known elements that were not radioactive. Earlier experience with the rare-earth elements had shown that it was often very difficult to separate chemically elements that were in fact distinct, and one might have been tempted to conclude that the radio-elements would eventually be separated and recognised as being chemically distinct. However, by this time the periodic table of Mendeleev (1869) was well established, and there were just not sufficient places to accommodate all the new elements that would have to be postulated. Thus Soddy was led to conclude (1911) that some of the radioelements were chemically identical to elements that were not radioactive. That is to say, the atoms of a given chemical element are not necessarily all identical, but rather can exist in different physical forms, to which the name *isotopes* was given.

By 1914 it had become clear that the different isotopes of a given element were distinguished by their atomic masses. This was established by showing chemically that the lead formed as the stable end product of two different radioactive sources had different atomic weights. In the meantime, however, the structure of the atom had been shown to be radically different from what had been envisaged by J J Thomson, as we discuss in the next section.

§2 THE RUTHERFORD–BOHR NUCLEAR ATOM

In 1906 Rutherford, still at McGill, observed that when α particles passed through a thin metal foil they were sometimes deflected, or *scattered*. He realised that this must be due to interactions between the α particle and the atoms in the foil, and saw that in fact the α particle could be used as a tool for probing the structure of the atom. It was in 1909 that Rutherford, by then in Manchester, began the systematic realisation of this project in a series of experiments performed by Geiger and Marsden.

The set-up of these experiments is shown schematically in figure 2.1. A natural source of α radioactivity was placed behind a lead screen with a small hole in it. The beam of α particles emerging from this collimator then passed through a metallic foil and impinged on a movable zinc sulphide screen, where the arrival of individual α particles was marked by tiny flashes of light. In this way the number of particles scattered through different angles by the foil could be counted†. Scattering of α particles by the foil was to be expected on account of the electrical fields that must exist even within the neutral atoms of the foil because of the discrete nature of the negatively charged electron (the α particle itself was already known to be a positively charged helium ion). However, according to the 'plum-pudding' model of the atom these fields should be rather weak, since even the local deviations from strict neutrality would have to be small. Thus it was expected that the beam of α particles would be only weakly scattered.

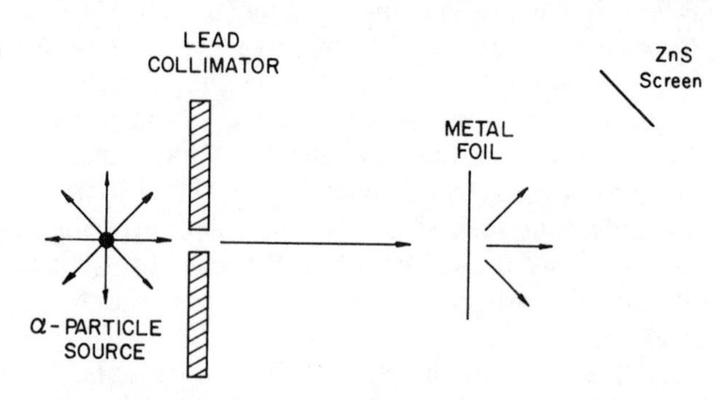

Figure 2.1 Set up of the Geiger–Marsden experiment.

This was indeed found to be the case for most of the α particles, but most unexpectedly some were scattered through quite large angles, sometimes even in excess of 90°. Rutherford recorded his astonishment thus: 'It was about as credible as if you had fired a 15-inch shell at a piece of tissue paper and it came back and hit you'. Such large deviations could not possibly arise from an encounter with a 'plum-pudding' atom: the required fields were just too large. Rutherford now saw that much higher fields (the only forces he considered were electrostatic) would arise if the positive and negative charges of the atom were separated. Accordingly he came to propose that there was a positive charge concentrated in a very small region at the centre of the atom, which became known as the *nucleus* of the atom. Overall neutrality of the atom was ensured by a cloud of electrons at a relatively great distance from the nucleus; the atomic radius of around 10^{-8} cm was essentially the radius of this cloud. Clearly, the positive charge on the

†This experiment, involving the scattering of beams of particles, was the prototype of what is still the most powerful technique in nuclear and elementary particle physics.

nucleus must be equal in magnitude to an integral multiple of the electron charge e: it will be equal to Ze where Z is the number of extranuclear electrons in the neutral atom. As will be discussed below, measurement of Z showed it to be quite a small number, of the order of half the atomic weight relative to hydrogen. This meant that most of the mass of the atom must reside in the nucleus, the electron being known to be nearly 2000 times lighter than the hydrogen atom.

Rutherford scattering law. Rutherford's analysis of the observed α scattering in terms of his nuclear model of the atom was based on the assumption that both nuclei, i.e. the target nucleus and the α particle, could be considered as point charges. The force law was thus taken to be Coulombic

$$V(r) = \frac{Z_1 Z_2 e^2}{r} \tag{2.1}$$

where Z_1 and Z_2 are the charges in e units of the target nucleus and α particle, respectively, and r the distance between these two nuclei.

According to classical mechanics, the only mechanics available at the time, the trajectory of the α particle relative to the target nucleus must be a hyperbola† the equation of which, in polar coordinates, can be written as

$$\frac{1}{r} = \frac{1}{a(\varepsilon^2 - 1)} (\varepsilon \cos \phi - 1). \tag{2.2}$$

Referring to figure 2.2, we are taking the target nucleus as origin O, and OA as the axis, A being the point of closest approach to O. Also, a is the distance AC, where C is the point of intersection of the asymptotes with OA, while ε is the eccentricity; as a general property of hyperbolas we have $\mathrm{OC} = \varepsilon a$.

If we regard the target nucleus O as fixed the net scattering angle of the α particle will be

$$\theta = \pi - 2\alpha \tag{2.3}$$

where α is the angle of inclination of each asymptote to the axis OA. But this latter angle is just the limiting value of ϕ for $r \to \infty$, so that from equation (2.2) we have

$$\cos \alpha = \frac{1}{\varepsilon} \tag{2.4}$$

from which

$$\theta = 2 \sin^{-1} \frac{1}{\varepsilon}. \tag{2.5}$$

†See, for example, Symon K R 1964 *Mechanics* 2nd edn (Reading, MA: Addison-Wesley) ch. 3.

In reality, of course, the target nucleus does not remain fixed but rather recoils in such a way that the total momentum of the two-body system is conserved. Thus the scattering angle θ of the relative motion, given by equation (2.5), is not the scattering angle measured by an observer in a laboratory-fixed frame. However, a simple relationship does exist between the two angles (see Appendix A).

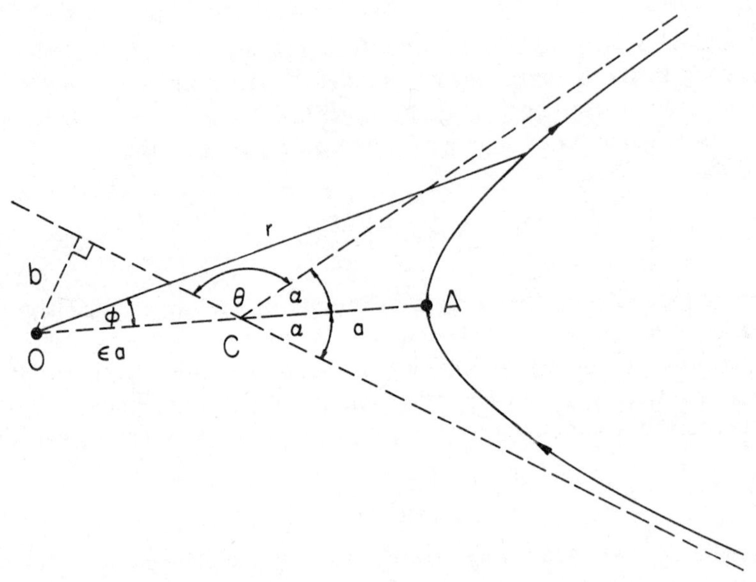

Figure 2.2 Trajectory for Rutherford scattering.

The eccentricity ε that we need in order to use equation (2.5) is determined by the total energy E_{CM} in a frame in which the centre of mass is at rest, and by the total orbital angular momentum l in this same frame. For the former of these two conserved quantities we have from equation (A.25)

$$E_{CM} = \tfrac{1}{2}\mu u^2 \tag{2.6}$$

where μ is the reduced mass of the two-particle system, given by equation (A.13) as

$$\mu = \frac{M_1 M_2}{M_1 + M_2} \tag{2.7}$$

and u is the asymptotic relative velocity. As for the total orbital angular momentum in the same frame of reference, equation (A.31) gives

$$\begin{aligned} l &= \mu u b \\ &= b\sqrt{2\mu E_{CM}} \end{aligned} \tag{2.8}$$

where b is the *impact parameter*, the distance of the target nucleus from the initial direction of approach of the projectile (see figure 2.2). A straightforward calculation, the details of which are left to the reader then gives (see Problem I.4)

$$\varepsilon=\left[1+\left(\frac{2E_{\mathrm{CM}}b}{Z_1Z_2e^2}\right)^2\right]^{1/2}. \tag{2.9}$$

Equation (2.5) for the scattering angle now becomes

$$\sin\frac{\theta}{2}=\left[1+\left(\frac{2E_{\mathrm{CM}}b}{Z_1Z_2e^2}\right)^2\right]^{-1/2}. \tag{2.10}$$

from which

$$\tan\frac{\theta}{2}=\frac{Z_1Z_2e^2}{2E_{\mathrm{CM}}b}. \tag{2.11}$$

In the usual scattering experiment, such as that of Geiger and Marsden, one does not know the impact parameter b for a given projectile; in fact all values of b, and hence all deflections, will be possible. We proceed as follows. Suppose first that there is only one target nucleus. Then if j is the flux density of the beam, i.e. the number of particles crossing unit area per unit time, the number of projectiles incident on the target per second with an impact parameter between b and $b+\mathrm{d}b$ is just

$$\mathrm{d}n(b)=j\,2\pi b\,\mathrm{d}b. \tag{2.12}$$

Substituting now from equation (2.11) for b we find for the number of projectiles scattered per second through an angle lying between θ and $\theta+\mathrm{d}\theta$

$$\mathrm{d}n(\theta)=2\pi j\frac{Z_1^2Z_2^2e^4}{8E_{\mathrm{CM}}^2}\cot\frac{\theta}{2}\operatorname{cosec}^2\frac{\theta}{2}\,\mathrm{d}\theta. \tag{2.13}$$

In reality, of course, there will be more than one target nucleus (see p. 123). Let there be N target nuclei per unit area of the target foil, and let s be the cross sectional area of the beam (we are supposing the target area to be greater than the cross section of the beam). Then assuming no multiple scattering the expression (2.13) for $\mathrm{d}n(\theta)$ will simply have to be multiplied by Ns

$$\mathrm{d}n(\theta)=2\pi IN\frac{Z_1^2Z_2^2e^4}{8E_{\mathrm{CM}}^2}\cot\frac{\theta}{2}\operatorname{cosec}^2\frac{\theta}{2}\,\mathrm{d}\theta \tag{2.14}$$

where

$$I=js \tag{2.15}$$

is the beam current.

Since the scattering angles lying between θ and $\theta+\mathrm{d}\theta$ define a solid angle

$$\mathrm{d}\Omega=2\pi\sin\theta\,\mathrm{d}\theta \tag{2.16}$$

we have for the rate of scattering per unit solid angle, which is the quantity that determines the actual counting rate of a given detector†,

$$\frac{dn}{d\Omega} = IN \frac{Z_1^2 Z_2^2 e^4}{16E_{CM}^2} \operatorname{cosec}^4 \frac{\theta}{2}. \tag{2.17}$$

This is the famous *Rutherford scattering law*, and its quantitative verification by Geiger and Marsden constituted a conclusive test of Rutherford's hypothesis of the nuclear atom.

Actually, what was verified in these experiments was the $\operatorname{cosec}^4 \theta/2$ form of the angular distribution. Measurement of the multiplicative constant in equation (2.17) led to a rough determination of the unknown charge number Z_1 of the target nucleus: it had already been established that the charge number of the α nucleus was $Z_2 = 2$ (actually, it was the α *particle* that had been shown to carry two units of charge, and the tacit assumption was made that it was completely ionised, i.e. consisted of a bare nucleus). The value obtained in the original experiments on gold was $Z_1 \simeq 100$, which is to be compared with the correct value of 79: the determination of this is discussed below. More generally, Geiger and Marsden showed that the charge number of the nucleus was roughly one half of the atomic weight, relative to the hydrogen atom.

The question now arises as to the size of the nucleus. The above analysis of Rutherford assumed point nuclei, interacting only through the Coulomb law (2.1), but it was realised that it would still be valid for finite nuclei, provided there was no interpenetration of the charge distributions, the field outside a finite charge distribution (spherically symmetrical) being Coulombic. Now the closest distance of approach r_{min} of the α particle with the target nucleus, corresponding to a head-on collision, is given by (see Problem I.5)

$$E_{CM} = V(r_{min})$$

i.e.

$$r_{min} = \frac{2Ze^2}{E_{CM}}. \tag{2.18}$$

Thus as long as the bombarding energy is low enough for r_{min} to be greater than the sum of the radii of the two charge distributions there will be no deviation from the Coulomb law and the Rutherford scattering relation (2.17) should be satisfied.

Now in the first experiments of Geiger and Marsden, performed on gold with α particles of energy less than 10 MeV, no deviations from the Rutherford scattering law could be found. With Z being known to be of the order of 100, Rutherford was able to infer from equation (2.18) that the sum of

†This will be proportional to $(dn/d\Omega)\ S/D^2$, where S is the exposed surface area of the detector and D its distance from the target.

the radii of the gold and α particle nuclei must be of the order of 10^{-12} cm or less. (Using units that are particularly convenient for nuclear physics, we have $e^2 = 1.44$ MeV fm, where we have introduced the fermi or femtometre: 1 fm $= 10^{-15}$ m.) As will be discussed in §4, experiments performed either with higher energy probes or on lighter targets did indeed show significant deviations from the Rutherford scattering law, thereby providing more precise information as to nuclear sizes.

It is appropriate at this point to remark on one other aspect of Rutherford's analysis of α particle scattering. As we have already pointed out, he was obliged to use classical mechanics, quantum mechanics still being some fifteen years away in the future. Is is a most fortunate circumstance that the one force for which classical mechanics and quantum mechanics agree is the Coulomb force. If the electrostatic force between two charges had taken any other form Rutherford's use of classical mechanics would have led to incorrect results, and the scattering experiments would have been incomprehensible. The implications of this would have been very serious indeed. Not only would the discovery of the nucleus have been compromised but so also would the development of quantum mechanics, since this depended so much on Rutherford's picture of the atom (see the discussion below on Bohr's quantum theory). In other words, the development of quantum mechanics might very well have been delayed by the manifestation of specifically quantum-mechanical effects!

Subsequent developments. The advent of Rutherford's nuclear model of the atom had the most profound implications for both physics and chemistry. When two atoms encounter each other under normal conditions, i.e. when they are moving with just their thermal energy, their respective nuclei will never come close enough for their mutual interaction to induce any changes in them. On the other hand the loose electron clouds surrounding the nuclei will be free to be deformed under their mutual influence, and there will be the possibility of some electrons passing from one atom to the other. Thus it came to be realised that the chemical properties of an atom would be determined entirely by the cloud of electrons, with the nucleus having no active role in chemical change at all (except insofar as mass is concerned, and insofar as the nuclear charge determines the configuration of the electron cloud). The chemical properties of an element would thus be determined essentially by the number, Z, of extranuclear electrons in the atom, and it thus became very tempting to identify this number Z, which also corresponds to the nuclear charge, with the *atomic number* of the element, i.e. the number determining the position of the element in the periodic table. This hypothesis was rapidly confirmed, as we shall discuss below.

Radioactive changes, on the other hand, being completely insensitive to external physical conditions such as temperature and pressure, were recognised as taking place in the nucleus rather than in the surrounding electron cloud. The accompanying transmutation of the chemical element could be attributed to a change in the nuclear charge.

Bohr's theory. Although classical mechanics had been instrumental in inferring the existence of the nucleus from the α scattering experiments of Geiger and Marsden, it was soon realised that the stability of the nuclear atom was completely incompatible with classical physics. If the extranuclear electrons were stationary then they would simply fall into the nucleus under the attraction of its positive charge. On the other hand, if the electrons moved around the nucleus in the same manner as the planets move around the sun, then, by virtue of the central acceleration, they should rapidly lose energy through emission of electromagnetic radiation and spiral into the nucleus.

Bohr was visiting Rutherford's laboratory in Manchester at the time of the discovery of the nucleus, and applied the new quantum concepts of Planck and Einstein to this problem of the stability of the nuclear atom. We shall assume that the reader is familiar with Bohr's quantum theory (1913), and recall simply that he postulated, in defiance of the laws of classical physics, the existence of certain stable electron orbits, characterised by angular momentum equal to an integral multiple of $\hbar = h/2\pi$, where h is Planck's constant. No other orbits were possible, and if an electron left one orbit it had to jump discontinuously to another allowed orbit, emitting (or absorbing) the energy difference as a photon.

This model was extremely successful when applied to the hydrogen atom, giving a precise rendering of the measured line spectrum: the energies of the permitted orbits were expressed as

$$E_n = -\frac{Z^2 e^4 m_e}{2\hbar^2}\frac{1}{n^2} \tag{2.19}$$

where n is some integer, and their radii (assuming circular orbits) as

$$r_n = \frac{\hbar^2}{Ze^2 m_e} n^2 . \tag{2.20}$$

Nevertheless, there were serious limitations to the model. Most notably it could not be extended systematically to multielectron atoms, and it was quite incapable of explaining the observed intensities of the different spectral lines, i.e. of the rates of transition of an electron from one orbit to another. Both these difficulties sprang from the basic conceptual difficulty that the Bohr theory consisted of a completely arbitrary mixture of classical and non-classical ideas, and lacked any coherent basis. Only with the advent of quantum mechanics in 1925 was the necessary complete revision of basic ideas forthcoming.

Now although the optical spectra of complex atoms certainly could not be analysed by the Bohr theory, it was observed that x-ray spectra bore a similarity to that of the hydrogen atom. This is because only the innermost orbits are involved, and since they are very little perturbed by electrons in other orbits, their energies will be given by the hydrogenic expression (2.19). It follows that a measurement of the x-ray spectrum of any element should allow a direct determination of Z. Such experiments were performed by Moseley (1913) on a large number of elements, and he found that the measured value Z of the nuclear charge could indeed be identified with the atomic number as determined by the position of the element in question in the periodic table. In principle, Z is determined even more directly by α scattering experiments of the type performed by Geiger and Marsden, but it was only in 1920 that their accuracy had been improved (by Chadwick) to the point where correct values could be obtained.

§3 CONSTITUENTS OF THE NUCLEUS

The lightest nucleus is that of the hydrogen atom, $Z = 1$, and it is known as the *proton*. It would be natural to suppose at first that a nucleus of atomic number Z consisted simply of Z protons, but this will not do, since the atomic weight would then be expected to be just Z times that of hydrogen, while in reality it is usually at least double this. Prior to 1932 the only way out of this dilemma was to suppose that there were electrons in the nucleus as well: there would be A protons and $A - Z$ electrons, leaving a net positive charge of Ze, and with the integer A determining the mass of the nucleus, since the electron's mass is very small compared with that of the proton.

Some confirmation of this picture became available after 1920 from the direct determination of nuclear masses made possible by Aston's *mass spectrograph*, which measured essentially the deflection of positive ions under the simultaneous influence of electric and magnetic fields. It was found that, at least for light nuclei, the nuclear mass was close to an integral multiple of the proton mass: the corresponding integer was then identified with the number of protons A. Nevertheless, with the advent of quantum mechanics in 1925 serious objections to the presence of electrons in the nucleus arose. One of these arguments is based on the Heisenberg uncertainty principle and runs as follows. If an electron is confined within a nucleus the uncertainty in its position is of the order of the nuclear diameter $2R$. There will then be an uncertainty in its momentum given by

$$\Delta p \sim \frac{\hbar}{2R} = \frac{\lambda\!\!\!^{-}_{ce}}{2R} m_e c \tag{3.1}$$

where

$$\lambda\!\!\!^{-}_{ce} \equiv \frac{\hbar}{m_e c} = 3.86 \times 10^{-13}\ \text{m} \tag{3.2}$$

is the Compton wavelength of the electron.

It follows that an electron confined to the nucleus must sometimes have a momentum of at least this magnitude, i.e. an energy of

$$E \simeq c\Delta p = \frac{\lambda\!\!\!^{-}_{ce}}{2R}\, m_e c^2 . \tag{3.3}$$

Now we have seen in the last section that R is always less than 10 fm, so that an electron confined to the nucleus must often have energies of the order of at least 15 MeV (this result justifies the use of the relativistic assumption in the first step of equation (3.3)). The question then arises as to how electrons of such high energy could be held within the nucleus. Actually, some nuclei *do* emit electrons spontaneously—this is β radioactivity (see §11)—but at the time energies as high as 15 MeV had never been encountered. And in any case, the fact remained that most of the nuclei known at the time did not emit electrons at all.

Another argument against the presence of electrons in the nucleus, which in fact was really decisive, came from the measured spin of the nitrogen nucleus. Discussion of this will be postponed until §6.

Neutron. The resolution of the difficulty came in 1932 with Chadwick's discovery of the long-suspected *neutron*, a neutral particle having a mass very close to that of the proton. As we shall often see later, the two particles have some other similarities as well, and are often referred to jointly as *nucleons*.

Heisenberg immediately proposed that the nucleus consisted exclusively of protons and neutrons, with no electrons present, so that in particular a nucleus of atomic number Z will contain Z protons. Also, if there are N neutrons in the nucleus, then the total number of nucleons

$$A = N + Z \tag{3.4}$$

known as the *mass number*, will *roughly* determine the mass of the nucleus, expressed in terms of the nucleon mass (see §5 for more details).

Aston had found in his mass-spectrograph measurements that there were often several different nuclear masses corresponding to the same chemical element. This was a direct manifestation of the isotopy phenomenon discovered in connection with the heavy radioactive elements and their stable end products (see §1). Aston, however, showed that the phenomenon was prevalent throughout the periodic table, and we now know that all elements can exist in several isotopic forms, although not all are stable (see below).

The nuclei of different isotopes of the same element will be distinguished by the number of neutrons, N; they will all have the same number of protons, Z. However, the accepted notation is that one identifies the different isotopes by the mass number A rather than by N, e.g. ^{1}H, ^{2}H, ^{207}Pb, ^{208}Pb, etc†.

The chemical symbol determines, of course, the value of Z, but it may be displayed explicitly, e.g. $^{1}_{1}H$, $^{2}_{1}H$, $^{207}_{82}Pb$, $^{208}_{82}Pb$, etc. Hydrogen is unique in that its different isotopes are often accorded distinct chemical symbols and names: ^{2}H is known as deuterium (D) and ^{3}H as tritium (T).

Different nuclei having the same mass number but different Z (and hence different N) are known as *isobars*. Clearly, different isobars having the same A will correspond to different chemical elements, so that the concept of isobar is of no interest to chemistry, although it is important for nuclear physics.

Binding of the nucleus. The question arises as to the nature of the forces binding the neutrons and protons together in a nucleus. Clearly they cannot be electromagnetic, since these forces will not act at all on the neutron (except very weakly through its magnetic moment) and between protons they are actually repulsive. We return to this question in §§5 and 8.

Range of known nuclides. The known chemical elements have Z ranging from 1 to just over 100, and, as we have already pointed out, all can exist in the form of several different isotopes. (The tendency is for N to be close to Z for light nuclei, but as we move to heavier nuclei there is a growing neutron excess, with N about 50% larger than Z for Z around 100 (see §14).) Altogether, about two thousand different *nuclides*, i.e. different nuclear species, are known at the present time, but only some 280 are completely stable, the rest transforming spontaneously, through α or β decay, into nuclei that are more stable, being more strongly bound. Actually, a large number of these radioactive nuclides occur naturally in the earth, many of them being sufficiently long-lived for appreciable quantities of them to have survived from the time they were originally synthesised inside some star‡. Others, while much shorter-lived, still occur naturally as the

†The old convention of writing H^{1}, H^{2}, Pb^{207}, Pb^{208}, etc, is no longer permitted, although one still says 'lead-207', for example. People who are used to writing 'five pounds', for example, as £5.00 should have no trouble.

‡See Chapter VI for an account of stellar nucleosynthesis. It suffices for the moment to realise that most nuclei heavier than hydrogen in the galactic dust out of which the solar system condensed were created in the interior of stars which subsequently blew apart in a supernova explosion (note, however, that an appreciable amount of helium and deuterium were made in the primordial 'big bang').

daughter products of longer-lived radioactive nuclides. Finally, we should note the formation of radioactive nuclides in the earth's atmosphere through nuclear reactions initiated by cosmic-ray bombardment; a famous example of this is ^{14}C (see §19).

However, most radioactive nuclides can only be obtained artificially (in nuclear reactors, by bombardment with an accelerator beam, or in the aftermath of a nuclear explosion), as the short-lived products of nuclear reactions, although it is known that many, if not all, must exist inside the stars at some stage or other of their evolution. (For this reason some of these unstable nuclides may have existed in the earth in the past, even if they are now 'extinct'.) As experimental techniques improve more and more highly unstable nuclei will be produced and identified in the laboratory.

While many elements have more than one stable isotope, others do not have any. This is particularly true of the heavy elements, and in fact bismuth ($Z = 83$) is the heaviest element to have any stable isotope. The general trend thereafter is that with increasing Z all nuclides become more and more short-lived, although thorium ($Z = 90$) and uranium ($Z = 92$) are exceptional in that both have isotopes sufficiently long-lived for them to occur naturally in the earth†. But uranium is the heaviest element that is found in a natural state in the earth, and the so-called *transuranic elements* can only be obtained artificially, although some are known to be formed in supernova explosions.

For some years now, a great effort has been devoted to the synthesis of elements of ever-increasing Z, mainly in Berkeley (USA), Dubna (USSR) and Darmstadt (W Germany). The method invariably followed for elements heavier than plutonium ($Z = 94$) consists of bombarding a suitable target with an accelerator beam of appropriate nuclei. At the time of writing (early 1985) $Z = 109$ represents the greatest atomic number for which the production of an isotope has been confirmed. For several years there was some theoretical speculation that if we could reach out as far as $Z = 114$ or so, an 'island of stability' might emerge, i.e. there may be some isotopes considerably more stable than those that have already been found in the vicinity of $Z = 100$. However, an extensive experimental search for these

†All the isotopes of all the other elements in this region, i.e. $Z = 84, 85, 86, 87, 88, 89$ and 91 have lifetimes that are much shorter than the age of the earth. Nevertheless, all these elements do occur naturally, one or more isotopes of each being formed as products of the radioactive decay of one or other of the long-lived thorium or uranium isotopes. As for the sub-bismuth region, $Z < 83$, the only elements that do not have any stable isotopes are technetium ($Z = 43$) and promethium ($Z = 61$). In fact, all their isotopes have lifetimes that are very short compared with the age of the earth, and since they are not formed as a product of natural radioactivity either, these elements are not found naturally in the earth: they are the only 'missing' elements below uranium.

'superheavy' elements has so far proven fruitless. (It is interesting to note that the exceptional stability of some of the thorium and uranium isotopes, compared with neighbouring nuclides, could be regarded as defining an island of stability.)

As for the mass numbers A, the range of these is limited by the range of the atomic numbers Z. The largest value of A that has been found so far is 267, associated with element 109. The heaviest *stable* nuclide is ^{209}Bi. Mass numbers $A = 5$ and 8 are remarkable in that they have no stable nuclide associated with them, a fact of great significance for stellar nucleosynthesis (see §28).

§4 SIZE OF THE NUCLEUS

It will be recalled from §2 that the early α scattering experiments of Geiger and Marsden showed no deviation from the Rutherford scattering law (equation (2.17)) and were consistent with point nuclei; more precisely, they showed that the nuclear radius was less than 10 fm. Deviations from the Rutherford law were first observed in experiments involving the scattering of α particles by hydrogen, performed in Rutherford's laboratory in 1919. They indicated a breakdown of the Coulomb law for distances of approach of less than 4 fm.

The interpretation of these experiments was complicated by two factors. Firstly, as soon as deviations from the Coulomb force law set in, classical mechanics breaks down and the use of quantum mechanics becomes essential. Secondly, the deviations from the Coulomb law could not be accounted for entirely by the finite extension of the nuclear charge distribution. That is to say, the interaction between the two nuclei was no longer purely electrostatic; rather, the specifically nuclear force responsible for the binding of nucleons in the nucleus reveals itself also in scattering. This complicating factor of the nuclear force, which is still not completely determined (see §8), does not arise when electrons are used as the probe, since in that case the interaction, being purely electromagnetic, is known to a very high degree of reliability. On the other hand, it must be remembered that electrons will interact only with the protons†, and therefore do not tell us anything about the neutron distribution.

†This is not altogether true: electrons are scattered weakly by the neutron's magnetic moment, and allowance for this has to be made in determining the proton distribution. Also, electrons of energy greater than 500 MeV have a de Broglie wavelength significantly smaller than the dimensions of the nucleon, and thus respond to the distribution of charge and currents within the individual protons and neutrons.

The technique of electron scattering has, in fact, been refined to a very high degree, and it has enabled the proton distribution to be determined in considerable detail. For our purposes it suffices to know that the general form of the proton distribution is as shown in figure 4.1. That is to say, the charge density ρ_{ch} is fairly constant over the bulk of the nucleus and falls off rapidly to zero in the surface. This distribution is well represented by a so-called Fermi function

$$\rho_{ch}(r) = \frac{\rho_0}{1 + \exp[(r - R)/a]} \tag{4.1}$$

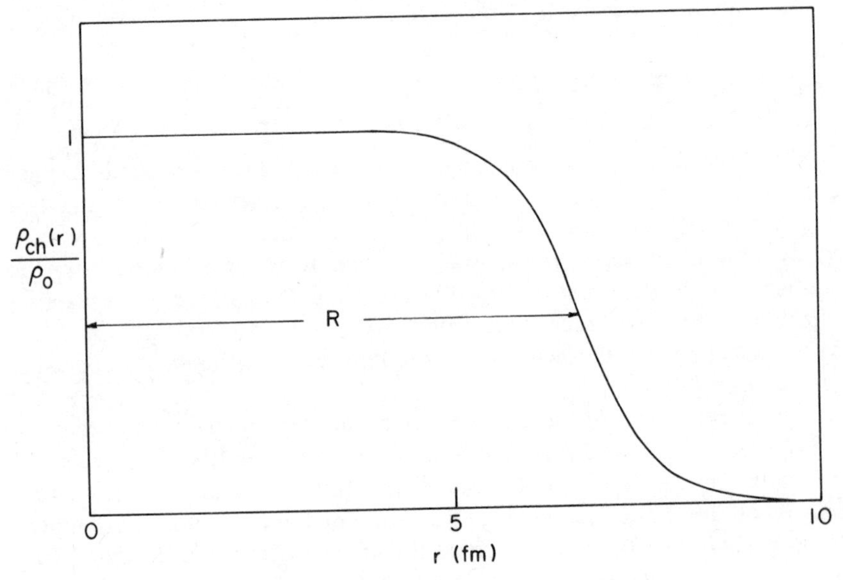

Figure 4.1 Fermi charge distribution in nucleus $A = 200$, with $r_0 = 1.15$ fm and $a = 0.5$ fm.

where

$$R \simeq r_0 A^{1/3} \tag{4.2}$$

with r_0 lying between 1.1 and 1.2 fm, and $a \simeq 0.5$ fm for all nuclei. With these values of r_0 and a it follows that for all nuclei the central density $\rho_{ch}(r = 0)$ will be very close to ρ_0. Thus the significance of R is that it represents the value of r for which the charge density has fallen to one half of its central value. The parameter a, on the other hand, gives a measure of the distance over which this fall-off takes place, i.e. it represents the surface thickness, or diffuseness. Since a is very much smaller than R for all but the lightest nuclei we can regard R as an effective nuclear radius†. Finally, ρ_0 is not really an independent parameter but is determined by the normalisation condition

$$4\pi \int_0^\infty \rho_{\mathrm{ch}}(r) r^2 \, \mathrm{d}r = Z. \tag{4.3}$$

As for the distribution of neutrons, this can only be measured by the scattering of particles that respond to the specifically nuclear forces, e.g. neutrons, protons, α particles and pions (see §8). Because of the residual uncertainties in this interaction it is inevitable that the neutron distribution cannot be measured with the same precision as the proton distribution. Nevertheless, it is clear that there are no significant differences, and that the Fermi function (4.1), with the same values of the parameters, serves equally well for protons and neutrons.

The relation (4.2) for the nuclear radius, with r_0 taking roughly the same value for all nuclei, means simply that the volume of a nucleus is proportional to A, its number of nucleons. That is, all nuclei have about the same density. With r_0 having the above value, this constant density lies between 1.5 and 1.6×10^{38} nucleons/cm^3, i.e. about 2.6×10^{11} kg cm^{-3}, an enormous figure which would correspond to the earth having shrunk to a diameter of about 1 km. Prior to 1968, it was believed that such densities just could not prevail on the macroscopic scale, but we now know otherwise: in §15 we shall see that certain stars can be regarded as gigantic nuclei.

This property of constant density displayed by nuclei is also exhibited by ordinary matter, the density of a sample being independent of its size (except on the astronomical scale; see §15). On the other hand, atoms do not show this property at all: all atoms have more or less the same *radius*. The constant density of nuclei immediately tells us something about the internucleonic force. This force is obviously attractive on the average, since otherwise nucleons would never be bound together in nuclei, but it cannot be *purely* attractive, since if it were the nucleus would collapse until all nucleons were within range of one another, in which case all nuclei would have roughly the same radius (as in the case of atoms). Rather, we may infer that the collapse is prevented by virtue of the internucleonic force becoming repulsive for very short distances, in exactly the same way that the constant density of ordinary matter is a consequence of a short-range repulsion in the interatomic, or intermolecular, force (see figure 12.2). We return to this question in §12.

†A supplementary source of information on the charge distribution in nuclei is given by the spectra of *muonic atoms*: see Problem I.14.

§5 BINDING ENERGY AND MASS OF THE NUCLEUS

If we are to break up a nucleus into its component nucleons and separate them completely from each other it will be necessary to supply energy to break the attractive bonds that hold the nucleons together. The minimum value of the energy necessary to effect such a separation is called the *binding energy*, B; it corresponds to the case where the nucleons will all have zero kinetic energy when they are completely separated (throughout this discussion of the energy it is assumed that the centre of mass of the nucleus is at rest). With the potential energy likewise being zero when the nucleons are all separated beyond the range of their mutual interactions, it follows that the nucleus must originally have had negative net internal energy

$$E = -B. \tag{5.1}$$

We see in the same way that B must represent the energy that is *released* when the nucleus is assembled from its component nucleons, originally well separated and at rest.

The internal energy E of a nucleus of A nucleons is given as an eigenvalue of the Schrödinger equation

$$H\Psi(\boldsymbol{r}_1, \boldsymbol{r}_2, \ldots, \boldsymbol{r}_A) = E\Psi(\boldsymbol{r}_1, \boldsymbol{r}_2, \ldots, \boldsymbol{r}_A) \tag{5.2}$$

where H is the Hamiltonian of the nucleus and the eigenfunction $\Psi(\boldsymbol{r}_1, \boldsymbol{r}_2, \ldots, \boldsymbol{r}_A)$ is the wavefunction, $\boldsymbol{r}_i$ denoting the position of the ith nucleon, etc. We write

$$H = \hat{T} + \hat{V} \tag{5.3}$$

where $\hat{T}$ is the kinetic energy operator for nucleonic motion within the nucleus

$$\hat{T} \equiv \frac{1}{2M} \sum_{i=1}^{A} \hat{p}_i^2 \equiv -\frac{\hbar^2}{2M} \sum_{i=1}^{A} \nabla_i^2 \tag{5.4}$$

and $\hat{V}$ is the total potential energy operator for interaction between the nucleons

$$\hat{V} = \sum_{i>j}^{A} \sum_{j=1}^{A} V_{ij}. \tag{5.5}$$

Here M is the nucleon mass (the difference between the neutron and proton masses can be ignored for most purposes), $\boldsymbol{p}_i$ the momentum of the ith nucleon (in the frame of reference where the centre of mass of the nucleus is at rest) and V_{ij} the interaction potential between nucleons i and j†.

At this point, it should be realised that the nucleus, like the atom, can usually exist in many different bound states, characterised by different wavefunctions Ψ, and by different values of the internal energy E (and also of other observable quantities, as discussed in the next two sections); these different states correspond, in fact, to the different eigensolutions of the nuclear Schrödinger equation (5.2). Again as in the case of atoms, a nucleus in one state of excitation can undergo an electromagnetic transition to a state of lower excitation by emitting a photon, a process known as γ *decay* in the nuclear context. Conversely, a nucleus can be excited by photon absorption.

In this book we shall be concerned mainly with nuclei in the ground state, i.e. the state with the greatest possible binding energy B (or the lowest possible internal energy E), since this is the state in which the vast majority of nuclei are found. Note, however, that in nuclear reactions and in β decays the products are often formed in an excited state, although usually they will γ decay very rapidly to the ground state. Nevertheless, in exceptionally hot stars, a certain fraction of nuclei remain in an excited state, their excitation being maintained by the very high-energy photons of the black-body radiation. It should also be realised that the study of the properties of excited states is of the greatest importance for our understanding of nuclear structure and forces.

If we could solve the Schrödinger equation (5.2), not only would the eigenvalues E give us the energies of the different states, but from the corresponding eigensolutions Ψ we could extract all possible information concerning all other conceivable properties of these states, e.g. their magnetic dipole and electric quadrupole moments and the rate of electromagnetic transitions between the different states. Unfortunately, solving the Schrödinger equation for the nucleus poses the most enormous problems. In the first place, the nucleon–nucleon interaction which has to be inserted into the equation is not completely known (see §8). Secondly, even if we specify this interaction, very serious calculational difficulties arise for more than two nucleons, i.e. for any nucleus more complex than that of deuterium, and for nuclei with more than three or four nucleons it is quite impossible to solve (5.2) at the present time. We shall return to this problem in §12, and devote the rest of this section to discussing how nuclear binding energies can be determined experimentally.

†We are assuming in equation (5.5) that the potential energy can be expressed entirely as a sum of pair-wise terms. In fact, it is known now that we should also consider many-body terms, corresponding to the fact that the interactions between any two nucleons can be influenced by the presence of other nucleons. However, such effects in no way change our general arguments.

Nuclear masses. The total internal energy of a nucleus reveals itself as a measurable difference between the mass M_{nuc} of the nucleus and the sum total of the masses of its constituent nucleons: according to the famous Einstein relation expressing the mass–energy equivalence†, we have

$$\frac{E}{c^2} = M_{nuc} - NM_n - ZM_p \tag{5.6}$$

where M_n and M_p denote the masses of the neutron and proton, respectively. The values of these latter quantities correspond to $M_n c^2 = 939.571$ MeV and $M_p c^2 = 938.279$ MeV.

Now the mass that is measured in a mass spectrograph is that of a positive ion and thus contains the mass of the electrons in the ion, but rather than subtract this to get the true nuclear mass the convention is to tabulate the mass M_{at} of the neutral atom, i.e. to the measured mass of the ion we add the mass of the missing electrons. Insofar as

$$M_{at} = M_{nuc} + Zm_e \tag{5.7}$$

where m_e is the electron mass, we can rewrite equation (5.6) in terms of atomic masses, thus

$$\frac{E}{c^2} = M_{at} - NM_n - ZM_H \tag{5.8}$$

where M_H is the mass of the neutral hydrogen atom.

Actually, in writing equation (5.7) we are effectively neglecting the binding energy of the electrons, so that the E defined in equation (5.8) is not quite the same as that defined in equation (5.6), for it includes the total binding energy of the electrons and this lies well outside the limits of experimental error of modern mass measurements (note that the binding energy of the few missing valence electrons in the positive ion whose mass is being measured is quite negligible). Nevertheless, the convention that it is the atomic masses rather than the nuclear masses that are tabulated is experimentally appropriate in that the energy release in reactions, fission and decays (see Chapters III and IV) is determined by the former, rather than the latter, since these processes involve neutral atoms (possibly a few outer electrons of negligible binding energy will be missing).

However, if we really want the nuclear internal energy E_{nuc} defined by equation (5.6), rather than the atomic internal energy E_{at} defined by equation (5.8)‡, a correction for the total electron binding energy becomes

†It is often said that the Einstein formula $E = mc^2$ lies at the basis of nuclear energy, since this involves the conversion of mass into energy. But of course the same is true for conventional chemical sources of energy; the only unique feature in the case of nuclear energy is that this is the only form where the mass changes are large enough to be measurable.

‡Note that the energy defined by equation (5.8) is really $E_{at} - ZE_{at}$ (H) (see Problem I.8). The last term is, however, quite negligible, since E_{at} (H) = − 13.6 eV.

necessary in principle. A convenient theoretical estimate of this correction gives

$$\begin{aligned} E \equiv E_{\mathrm{nuc}} &= E_{\mathrm{at}} + 14.33 Z^{2.39}\mathrm{eV} \\ &= (M_{\mathrm{at}} - NM_{\mathrm{n}} - ZM_{\mathrm{H}})c^2 + 14.33\, Z^{2.39}\mathrm{eV}. \end{aligned} \tag{5.9}$$

For most purposes this correction is quite negligible. (Note, however, that since our estimate for this correction is not exact any tabulation of bare nuclear masses would be inherently less accurate than the conventional tabulation of atomic masses.)

Atomic massses are usually expressed in terms of the *atomic mass unit*, which is defined as 1/12 of the mass of the neutral atom of ^{12}C

$$\begin{aligned} 1\ \mathrm{AMU} &= 1.66057 \times 10^{-27}\ \mathrm{kg} \\ &= 931.501\ \mathrm{MeV}/c^2. \end{aligned}$$

Often, as in the mass table found at the end of this book, the atomic mass (in AMU) is given in terms of the *mass excess*

$$\Delta = (M_{\mathrm{at}} - A)c^2 \tag{5.10}$$

expressed in MeV; clearly, for ^{12}C we shall have $\Delta = 0$. In place of equation (5.8), we shall have

$$E = \Delta - N\Delta_{\mathrm{n}} - Z\Delta_{\mathrm{H}} \tag{5.11}$$

where the mass excesses of the neutron and proton are respectively

$$\Delta_{\mathrm{n}} = 8.07137\ \mathrm{MeV} \qquad \Delta_{\mathrm{H}} = 7.28903\ \mathrm{MeV}.$$

(The reason why the ^{12}C atom rather than the ^{1}H atom is taken as the basis for the definition of the AMU is that atomic masses so expressed are always very close to the mass number, i.e. the mass excess Δ is always very small.)

Measuring the mass of a nucleus is the only way of determining *absolutely* its internal energy. However, the measurement of decay energies (Chapter III) and of nuclear-reaction Q values (Chapter IV) will determine the *relative* values of the internal energy of different nuclei, and as such can be used to extend the mass table to nuclei whose masses cannot be measured directly in a spectrograph, e.g. unstable nuclei. (The mass of the neutron, which cannot of course be measured in a mass spectrograph, is determined from its β decay; see §16.)

The systematics of nuclear binding energies will be discussed in some detail in §14, and here we note simply that most nuclei have a binding energy of about 8 MeV per nucleon, a conspicuous exception being deuterium, for which the binding energy is 1.11 MeV per nucleon. This is to be compared with the typical atomic or molecular binding energy, which is, of course, electromagnetic in origin, and is of the order of a few eV. This shows that the forces binding neutrons and protons together must be very much stronger than electromagnetic forces, which we have already noted (end of §3) could not in any case be responsible for nuclear binding.

It is interesting at this point to examine the possible role of gravity in nuclear binding. Qualitatively this could work, since it acts on both neutrons and protons and is always attractive. However, the gravitational energy of a nucleus of mass number A, radius R, is $-\frac{3}{5}GA^2M^2/R$, and even for a heavy nucleus with $A \simeq 200$ this is only of the order of 10^{-26} eV. Thus gravity is utterly negligible in ordinary nuclei, although we shall see in §15 that there are nuclear situations where it can play a dominant role. The nature of the forces which actually bind ordinary nuclei is discussed in §8.

§6 ANGULAR MOMENTUM AND MAGNETIC MOMENT

One of the many similarities between the neutron and the proton is that they are both fermions, i.e. like electrons, they both obey the Pauli exclusion principle, and also have intrinsic angular momentum, or *spin*, of $\frac{1}{2}\hbar$. Let us now denote by S the resultant of the spins of all the A nucleons in a nucleus of mass number A. According to the quantum-mechanical rules of angular momentum, the magnitude of $\boldsymbol{S}$ must be an even or odd multiple of $\frac{1}{2}\hbar$, according to whether A is even or odd, respectively.

In addition to their intrinsic angular momenta, the nucleons of a nucleus also have an orbital angular momentum associated with their motion within the nucleus. The resultant, $\boldsymbol{L}$, of the orbital angular momenta of all the nucleons of a nucleus must have a magnitude which is an integral multiple of $\hbar$, regardless of whether A is even or odd.

The total angular momentum of the nucleus is now

$$\boldsymbol{J} = \boldsymbol{L} + \boldsymbol{S}. \tag{6.1}$$

This is a rigorously conserved quantity, and it is the only angular momentum of the nucleus that can be directly measured (we indicate below how this can be done). Often J is referred to loosely as the 'spin' of the nucleus, although this can give rise to some confusion. Note particularly that different states of a given nucleus can have different values of J. In view of what has been said about $\boldsymbol{L}$ and $\boldsymbol{S}$ the magnitude of $\boldsymbol{J}$ must be an even or odd multiple of $\frac{1}{2}\hbar$, i.e. integral or half-integral, as A is even or odd, respectively.

This last remark was of crucial importance for the historical development of nuclear physics. Consider the nucleus ^{14}N. Since we now know that it contains 14 nucleons (7 protons and 7 neutrons) it must have integral angular momentum J, in units of $\hbar$. But according to the proton–electron model of the nucleus it contains 14 protons and 7 electrons, i.e. an odd number of fermions, so that J should be half-integral. In 1928 measurements of the hyperfine structure (see below) of the atomic spectrum of ^{14}N showed that $J = 1$ for this nucleus, thereby giving conclusive evidence against the existence of electrons in nuclei.

A magnetic moment $\boldsymbol{\mu}_l$ will be associated with the total orbital angular momentum $\boldsymbol{L}_p$ of the protons (but not with that of the neutrons, of course). With $\boldsymbol{L}_p$ expressed in units of $\hbar$, both classical and quantum mechanics give

$$\boldsymbol{\mu}_l = \frac{e\hbar}{2M_p c} \boldsymbol{L}_p \tag{6.2}$$

a result which is well confirmed experimentally. At the same time, both the neutron and the proton are found to have intrinsic magnetic moments parallel to their spins. It is convenient to express the measured values of these quantities in terms of the *gyromagnetic ratios* g_n and g_p, defined according to

$$\boldsymbol{\mu}_n = \frac{e\hbar}{2M_n c} g_n \boldsymbol{s}_n \tag{6.3a}$$

$$\boldsymbol{\mu}_p = \frac{e\hbar}{2M_p c} g_p \boldsymbol{s}_p \tag{6.3b}$$

respectively. The measured values are $g_n = -3.826\,29$ and $g_p = 5.585\,50$. The deviation of the former from zero and of the latter from unity cannot be understood in classical terms at all†. In fact, these values give significant information on the structure of the neutron and proton.

The resultant, observed magnetic moment of a nucleus can clearly be written as

$$\boldsymbol{\mu} = \frac{e\hbar}{2M_p c}(\boldsymbol{L}_p + g_p \boldsymbol{S}_p) + \frac{e\hbar}{2M_n c} g_n \boldsymbol{S}_n \tag{6.4}$$

where $\boldsymbol{S}_p$ and $\boldsymbol{S}_n$ are the total spins of the protons and of the neutrons, respectively, with

$$\boldsymbol{S}_p + \boldsymbol{S}_n = \boldsymbol{S}. \tag{6.5}$$

It is quite evident that μ will not be determined uniquely by J, since it depends on the way $\boldsymbol{J}$ is distributed between $\boldsymbol{L}$ and $\boldsymbol{S}$, and also between neutrons and protons. For precisely this reason the measured values of μ provide valuable information over and above that carried by J‡.

†The electron, too, has an anomalous magnetic moment corresponding to a gyromagnetic ratio of 2.002 393 16 rather than the classical value of unity. This result is well accounted for by quantum electrodynamics. See, for example, §6.6 of the book by Frauenfelder and Henley quoted in the General Bibliography.

‡See, for example, §5.6 of the book by Elton quoted in the General Bibliography.

We see furthermore that the magnetic moment $\boldsymbol{\mu}$ of a nucleus will not even be parallel to its angular momentum $\boldsymbol{J}$ in general. Nevertheless, for a given nuclear state, the angle between $\boldsymbol{\mu}$ and $\boldsymbol{J}$ will be fixed, classically speaking. Thus, like the vector $\boldsymbol{J}$, the magnetic moment $\boldsymbol{\mu}$ will, according to the usual quantum-mechanical rules of angular momentum, have $2J+1$ possible orientations with respect to any given direction. It follows that the energy of interaction between the nucleus and an externally applied magnetic field $\boldsymbol{H}$, given by

$$E_{\text{int}} = -\boldsymbol{\mu} \cdot \boldsymbol{H} \tag{6.6}$$

can take $2J+1$ values. This result forms the basis for the measurement of $\boldsymbol{\mu}$ and J†. The applied magnetic field may be generated artificially by magnets, or by the extranuclear atomic electrons. In the latter case, the interaction (6.6) manifests itself simply as hyperfine structure of the atomic spectra, the multiplicity of which determines J directly.

§7 PARITY

In addition to the angular momentum J, another quantity that characterises the different states of a nucleus is the *parity*, π. Unlike J, this can take only one of two values, + or −, as we now explain.

The parity operation P is the one of reflection of the axes of space co-ordinates: $(x, y, z) \rightarrow (-x, -y, -z)$. It implies changing from a right-handed to a left-handed system. For any nuclear wavefunction we shall have

$$P\Psi(\boldsymbol{r}_1, \boldsymbol{r}_2, \ldots, \boldsymbol{r}_A) = \Psi(-\boldsymbol{r}_1, -\boldsymbol{r}_2, \ldots, -\boldsymbol{r}_A) \tag{7.1}$$

Repeating the operation gives

$$P^2\Psi(\boldsymbol{r}_1, \boldsymbol{r}_2, \ldots, \boldsymbol{r}_A) = \Psi(\boldsymbol{r}_1, \boldsymbol{r}_2, \ldots, \boldsymbol{r}_A) \tag{7.2}$$

from which

$$P^2 = 1. \tag{7.3}$$

It follows that P can have only ± 1 as eigenvalues.

States which are eigenstates of P belonging to the eigenvalue $+1$ are said to be of *positive* or *even* parity, while those belonging to eigenvalue -1 are

† J can also be obtained from the spectra of diatomic molecules: see §2.5 of Elton's book.

said to have *negative* or *odd* parity. A given wavefunction may not have definite parity but it can always be expressed as a sum of functions having even and odd parity, respectively:

$$\Psi \equiv \tfrac{1}{2}(\Psi + P\Psi) + \tfrac{1}{2}(\Psi - P\Psi). \tag{7.4}$$

Let us now confine ourselves to wavefunctions that are eigenfunctions of the nuclear Hamiltonian

$$H\Psi = E\Psi \tag{7.5}$$

where H is given by equations (5.3), (5.4) and (5.5). Then

$$P(H\Psi) = P(E\Psi) = EP\Psi \tag{7.6}$$

from which

$$PHP^{-1}P\Psi = EP\Psi \tag{7.7}$$

which may be rewritten as

$$H'(P\Psi) = E(P\Psi) \tag{7.8}$$

where we have introduced the transformed Hamiltonian resulting from the parity operation

$$H' = PHP^{-1}. \tag{7.9}$$

Now to a very high degree of precision the nuclear Hamiltonian is found experimentally to be invariant under the parity operation

$$H' = H. \tag{7.10}$$

This will clearly be the case when the two-body potential terms V_{ij} appearing in equation (5.5) depend only on the distances between the two particles i and j. However, because of the weak interaction (see §11) equation (7.10) will not be satisfied exactly, i.e. the nuclear Hamiltonian is not strictly invariant under the parity operation: one says that 'parity is not conserved' (see Problem I.11). Still, the parity-violating effect of the weak interaction is very small, and we shall assume here that equation (7.10) holds.

Then equation (7.8) becomes

$$H(P\Psi) = E(P\Psi) \tag{7.11}$$

so that in view of equation (7.5) we can say that Ψ and $P\Psi$ are both eigenstates of H belonging to the same eigenvalue E. If now this eigenvalue is nondegenerate Ψ and $P\Psi$ must be simple multiples of each other

$$P\Psi = a\Psi. \tag{7.12}$$

But this just means that Ψ is an eigenstate of P, so that in this case Ψ must have definite parity, with the factor a being limited to the values ± 1.

If, on the other hand, the eigenvalue E is degenerate no such conclusion can be drawn. This occurs in the hydrogen atom, where the 2s and 1p states, for example, are degenerate, at least in the non-relativistic approximation. The former has positive parity and the latter negative, so that any linear combination of the two will not have definite parity, and yet will still be an eigenfunction of the Hamiltonian. However, because of the complexities of nuclear forces, no such exact degeneracies occur in nuclei, so that all nuclear states have well defined parity.

§8 NUCLEAR FORCES: THE STRONG INTERACTION

So far we have learned that the basic force between nucleons, the so-called nucleon–nucleon (N–N) force, is very much stronger than the electromagnetic interaction, but is also of very much shorter range, being quite negligible when the two nucleons in question are more than 2 or 3 fm apart. In order to emphasise this distinction from the electromagnetic interaction (and also from the gravitational interaction) one speaks of the *strong interaction*.

Most of our quantitative information on the N–N interaction comes from direct experiments on the two-nucleon system. These consist mainly of measurements on the scattering of proton and neutron beams by proton targets, which serve to elucidate the proton–proton (p–p) and the neutron–proton (n–p) interactions, respectively. Unfortunately, it is not possible to produce targets of free neutrons, so that *direct* measurement of the n–n interaction is not possible. Additional information on the n–p interaction comes from the single bound state formed by the n–p system: this is the *deuteron* (d), the deuterium nucleus. Neither the n–n nor the p–p systems form bound states.

There is, however, one piece of very valuable quantitative information on the nucleon–nucleon interaction that can be inferred from the properties of complex nuclei. Consider pairs of so-called *mirror nuclei*, e.g. ${}^{3}_{1}H_{2}$, ${}^{3}_{2}He_{1}$, or ${}^{23}_{11}Na_{12}$, ${}^{23}_{12}Mg_{11}$. Each member of any such pair can be transformed into the other by replacing *all* neutrons by protons, and vice versa. When allowance is made for the difference in Coulomb energy between the two members of any pair, the two nuclei are found to have essentially identical properties, both with respect to the ground state and also to all excited states (see §14 and Problem II.8). We can thus conclude that, to a high degree of precision, the n–n interaction is identical to the p–p interaction (it being understood that the Coulomb force is excluded from the latter). This quite general property of the nucleon–nucleon interaction is known as *charge symmetry*, and it implies that the inability to measure the n–n interaction directly is not a serious handicap.

Spin dependence of N–N interaction. It is reasonable to assume that in the ground state of the deuteron the two nucleons will have relative orbital angular momentum $l = 0$, since this will minimise the kinetic energy (see, however, the remarks below on the tensor force). Then according to equation (6.1) the total angular momentum of the deuteron will be

$$\boldsymbol{J} = \boldsymbol{S} \tag{8.1}$$

where

$$\boldsymbol{S} = \boldsymbol{s}_1 + \boldsymbol{s}_2 \tag{8.2}$$

is the resultant of the spins $\boldsymbol{s}_1$ and $\boldsymbol{s}_2$ of the two nucleons of the deuteron. Since the magnitude of S will be 0 or 1, in units of $\hbar$, it follows that the measured value of J must likewise be 0 or 1.

Now it turns out that the measured value of J is always 1, never 0, which can only mean that the n–p force is more attractive when the spins of the two nucleons are coupled together to give $S = 1$ than when they give $S = 0$ for the total spin. To see how such a spin dependence of the interaction might be represented we recall a standard result from the quantum-mechanical theory of angular momentum. Following Appendix B let

$$\boldsymbol{J} = (J_x, J_y, J_z) \tag{8.3}$$

denote any angular momentum. Then the operator $\hat{J}^2$ for the quantity J^2 has eigenvalues $J(J + 1)\,\hbar^2$

$$\hat{J}^2\psi_J = J(J + 1)\hbar^2\psi_J \tag{8.4}$$

where the quantum number J can be any integer or half integer.

Thus the operators $\hat{s}_1^2, \hat{s}_2^2$ for the individual nucleons have the single eigenvalue $\frac{1}{2}(\frac{1}{2} + 1)\hbar^2 = \frac{3}{4}\hbar^2$. On the other hand, the operator $\hat{S}^2$ for the pair of nucleons will have eigenvalue $1(1 + 1)\hbar^2 = 2\hbar^2$ for $S = 1$, and 0 for $S = 0$. Now from equation (8.2) we can write

$$\hat{S}^2 = \hat{s}_1^2 + \hat{s}_2^2 + 2\hat{s}_1 \cdot \hat{s}_2. \tag{8.5}$$

It follows then that the operator $\hat{s}_1 \cdot \hat{s}_2$ has eigenvalue $\frac{1}{4}\hbar^2$ for $S = 1$, and $-\frac{3}{4}\hbar^2$ for $S = 0$. The dependence of the n–p interaction on the total spin S can now be represented by writing it as

$$V = V_0(r) + V_1(r)\boldsymbol{s}_1 \cdot \boldsymbol{s}_2 \tag{8.6}$$

the contribution of the second term depending on whether $S = 0$ or 1.

Charge independence of nuclear forces. Having noted the charge symmetry property of the N–N interaction, i.e. the identity of the n–n and p–p forces, it is natural to conjecture that the n–p force might also be identical to these two. This more far-reaching hypothesis is known as *charge independence* and might appear at first sight to be ruled out by the following consideration: while the n–p system has a bound state the n–n does not (the absence of a bound p–p state is less conclusive, since it could always be a result of the Coulomb repulsion). However, in applying the hypothesis of charge independence we must remember that unlike the n–p system, the n–n and p–p systems must respect the Pauli principle. That is, the total wavefunction of a system of two like fermions is antisymmetric under exchange of all their coordinates, spatial and spin:

$$\Psi(1,2) = -\Psi(2,1). \tag{8.7}$$

If we factor this total wavefunction into a spatial part and a spin part, thus

$$\Psi(1,2) = \Phi(\boldsymbol{r}_1, \boldsymbol{r}_2)\chi(1,2) \tag{8.8}$$

where $\chi(1,2)$ represents the spin wavefunction of the two fermions, then the Pauli principle becomes

$$\Phi(\boldsymbol{r}_1, \boldsymbol{r}_2)\chi(1,2) = -\Phi(\boldsymbol{r}_2, \boldsymbol{r}_1)\chi(2,1). \tag{8.9}$$

Now we show in Appendix B that if the two spins couple together to give a total spin of $S = 1$ (triplet states) then the spin wavefunction is symmetric under exchange of the two fermions. On the other hand for $S = 0$ (singlet state) this function will be antisymmetric. Thus the Pauli principle (8.9) becomes

$$\Phi(\boldsymbol{r}_1, \boldsymbol{r}_2) = (-)^S \Phi(\boldsymbol{r}_2, \boldsymbol{r}_1). \tag{8.10}$$

Let us transform now to the system of coordinates

$$\boldsymbol{R} = \tfrac{1}{2}(\boldsymbol{r}_1 + \boldsymbol{r}_2) \tag{8.11a}$$

$$\boldsymbol{r} = \boldsymbol{r}_1 - \boldsymbol{r}_2 \tag{8.11b}$$

(see Appendix A), so that we can write $\Phi(\boldsymbol{r}_1, \boldsymbol{r}_2)$ in the form

$$\Phi(\boldsymbol{r}_1, \boldsymbol{r}_2) = F(\boldsymbol{R})\psi(\boldsymbol{r}). \tag{8.12}$$

The function depending on the centre of mass coordinate $\boldsymbol{R}$ is clearly symmetric under particle exchange, so that equation (8.10) reduces to the condition

$$\psi(\boldsymbol{r}) = (-)^S \psi(-\boldsymbol{r}) \tag{8.13}$$

on the relative motion.

Suppose now that the relative motion has well defined orbital angular momentum, so that we can write

$$\psi(\boldsymbol{r}) = R(r) Y_l^m(\theta, \phi) \tag{8.14}$$

where $R(r)$ is some radial function, and we have introduced the spherical harmonic

$$Y_l^m(\theta,\phi) = \mathrm{i}^{m+|m|}\left(\frac{2l+1}{4\pi}\frac{(l-|m|)!}{(l+|m|)!}\right)^{1/2} P_l^m(\cos\theta)\mathrm{e}^{\mathrm{i}m\phi}. \tag{8.15}$$

Now the operation $\boldsymbol{r} \to -\boldsymbol{r}$ is equivalent to $r \to r$, $\theta \to \pi - \theta$, $\phi \to \phi + \pi$. Then since the associated Legendre polynomial satisfies

$$P_l^m(-\cos\theta) = (-)^{l+m} P_l^m(\cos\theta) \tag{8.16}$$

it follows that the spherical harmonic transforms as

$$Y_l^m(\theta,\phi) \to (-)^l Y_l^m(\theta,\phi) \tag{8.17}$$

from which equation (8.14) leads to

$$\psi(-\boldsymbol{r}) = (-)^l \psi(\boldsymbol{r}). \tag{8.18}$$

Combining now equations (8.13) and (8.18) we find

$$(-)^{l+S} = 1 \tag{8.19}$$

i.e. the possible states of two identical fermions are limited by

$$\begin{aligned} S &= 1 \qquad l \text{ odd} \\ S &= 0 \qquad l \text{ even.} \end{aligned} \tag{8.20}$$

Odd values of l for $S = 0$ and even values for $S = 1$ are forbidden by the Pauli exclusion principle. However, while these restrictions apply to the n–n or p–p systems they do not apply to the n–p system. Thus the n–p system can certainly exist in states not satisfying the condition (8.20), i.e. $S = 1$, l even, and $S = 0$, l odd, but since the n–n and p–p systems cannot, no comparison of their interactions is possible for these states. Thus any meaningful discussion of the hypothesis of charge independence must of necessity be limited to states satisfying equation (8.20).

We now see why the absence of a bound di-neutron, far from being incompatible with charge independence, is actually required by it, since being $l = 0$ it would have to be $S = 0$, and we know that the n–p system is not bound in this state. Of course, while the non-existence of the di-neutron is a necessary condition for the validity of charge independence it is not sufficient: only careful and extensive scattering experiments afford a complete test. It is found that charge independence holds to a good approximation, but not as well as for charge symmetry.

Tensor force. It is important at this point to examine more carefully our earlier assumption that the deuteron is in an $l = 0$ state. If this were the case the charge distribution in the deuteron would be spherically symmetrical, but measurement of its electric field shows that this is not so. More specifically, the deuteron is found to have an electric *quadrupole moment*, and a quantitative analysis of this shows that whereas the $l=0$ state does indeed predominate there is also a weak admixture ($\sim 5\%$) of an $l=2$ state (it is easy to see that parity conservation prohibits a coupling of the $l=1$ state to the $l=0$ state).

This mixture of states of different orbital angular momentum can only mean that the N–N interaction contains a term that is non-central. The spins of the nucleons serve to mark out a preferred direction in space, and we must envisage a coupling between them and the vector $\boldsymbol{r}$, the join of the two nucleons. A constraint on the form of this coupling is imposed by the condition that although the orbital angular momentum $\boldsymbol{l}$ is not conserved the total angular momentum of the N–N system, $\boldsymbol{J}=\boldsymbol{l}+\boldsymbol{S}$, must be. It may be shown, then, that a possible form of this non-central term in the interaction is $V_T(r)\mathscr{S}_{12}$, where the radial function $V_T(r)$ can take any arbitrary form, while

$$\mathscr{S}_{12}=\frac{3}{r^2}(\hat{\boldsymbol{s}}_1\cdot\boldsymbol{r})(\hat{\boldsymbol{s}}_2\cdot\boldsymbol{r})-\hat{\boldsymbol{s}}_1\cdot\hat{\boldsymbol{s}}_2. \tag{8.21}$$

Such a non-central term is referred to as the *tensor force*. It has exactly the same form as the interaction between two small magnets, although it must not be thought that the tensor interaction between two nucleons is electromagnetic in origin.

The existence of the tensor force has been absolutely basic to the development of nuclear theory. However, it complicates matters considerably, and we shall not say much about it in this book, since many of its effects (for example, on the binding energy) can be simulated by an equivalent central force.

Pion exchanges and the nuclear force. It was proposed by Yukawa (1935) that, just as the electromagnetic interaction between two electric charges is mediated by the exchange of photons, the strong interaction between two nucleons was mediated by the exchange of some hitherto undiscovered particle. He showed that the form of the N–N interaction should vary asymptotically as

$$V(r)\sim\frac{f}{r}\exp(-r/\lambda\!\!\!^{-}_{c\pi}) \tag{8.22}$$

where f is some constant related to the strength of the interaction, while

$$\lambda\!\!\!^{-}_{c\pi}=\frac{\hbar}{m_\pi c} \tag{8.23}$$

in which m_π is the mass of the exchanged particle ($\lambda\!\!\!^{-}_{c\pi}$ is thus the Compton wavelength of this particle).

We see from equation (8.22) that $\lambda\!\!\!^{-}_{c\pi}$ is a measure of the range of the N–N force, since the smaller it is the more rapid the exponential fall-off. Thus the heavier the exchanged particle the shorter the range of the force. (We note that if the exchanged particle is massless, like the photon, then equation (8.22) reduces to the form of the Coulomb law.)

Particles that could be identified with those proposed by Yukawa were first discovered in cosmic rays by Powell (1947) and later in accelerator experiments. There are in fact three such particles, one of which is neutral, while the other two carry charge $\pm e$, respectively. They are known as *pi mesons*, or *pions*, and are denoted by π^0, $\pi^\pm$, respectively. All three are highly unstable, the π^0 decaying with a half-life of about 10^{-16} s, mainly into two photons, while the charged pions decay mainly into muons and μ neutrinos, with a half-life of the order of 10^{-8} s†. All three pions have zero intrinsic spin, i.e. they are bosons. Their masses are measured to be close to $270m_e$, so that $\lambda\!\!\!^{-}_{c\pi} \simeq 1.4$ fm, which is of the order of what we have already established for the range of the N–N force.

The inverse relationship between the mass of the exchanged particle, i.e. the pion, and the range of the N–N force can be understood very simply by means of an argument based on the Heisenberg uncertainty principle in the form

$$\Delta E\,\Delta t \sim \hbar. \tag{8.24}$$

In the present context this can be interpreted as stating that a violation ΔE of energy conservation can be tolerated for the indicated time Δt.

Suppose now that the two interacting nucleons are at rest. The emission of a pion by one nucleon will entail a violation of energy conservation

$$\Delta E = m_\pi c^2. \tag{8.25}$$

According to equation (8.24) the pion must be reabsorbed within a time

$$\Delta t = \frac{\hbar}{m_\pi c^2}. \tag{8.26}$$

It can do this either by falling back on the emitting nucleon or by encountering the second nucleon. But reabsorption by the second nucleon will only be possible if it is close enough to the first nucleon for the pion to reach it within the time given by equation (8.26). Making the most optimistic assumption that the pions can move with the velocity of light, we see that

†The muon is a particle that is very similar to the electron, except that it is some 207 times heavier, and is itself unstable, while the μ neutrino is similar to the electron neutrino discussed in §11, although distinct from it. See, for example, the book by Frauenfelder and Henley quoted in the General Bibliography.

the distance between the nucleons must be less than

$$d = c\,\Delta t = \frac{\hbar}{m_\pi c} = \lambda\!\!\!^{-}_{c\pi}. \tag{8.27}$$

If the separation of the nucleons exceeds this value then pions cannot be exchanged between them, in which case the two nucleons will not be able to interact with each other. But this means simply that d, as given by equation (8.27), must represent the range of the interaction.

Yukawa's theory was limited to the case where only one pion was exchanged at a time. However, we could envisage the exchange of two, three or more pions simultaneously. It is clear, however, in view of the above argument, that the greater the multiplicity of the pion exchange the shorter the range over which the corresponding mechanism is effective, simply because the exchanged mass will be greater. Thus the long-range part of the interaction will be dominated by single-pion exchanges (one refers to the 'Yukawa tail' of equation (8.22)), while increasingly complex multipion exchanges will become significant with decreasing separation of the two nucleons. The theory of the short-range part of the interaction (< 1 fm) has still not been worked out in all its details†.

Only high-energy (200–300 MeV) N–N scattering can determine the short-range part of the interaction. (The predictions of the pion exchange theory for the longer-range parts of the interaction have been well confirmed by scattering experiments.) It is found that the interaction does indeed become highly repulsive at distances less than about 1 fm, exactly as we had inferred from the fact that all nuclei have about the same density (§4).

In the above argument based on the Heisenberg uncertainty principle, it was assumed that the two interacting nucleons were at rest, so that the creation of a pion involved of necessity the violation of energy conservation. Since this violation can only be temporary, such pions must be reabsorbed, and hence cannot be observed by any detector; we speak therefore of *virtual* pions. In order to create real pions, i.e. pions that can actually be observed, it will be necessary for the relative energy of the two nucleons to exceed the pion rest-mass energy of 140 MeV. That is why pions can only be observed in cosmic rays or high-energy accelerator experiments (remember that the pion has a very short half-life).

†It is now believed that both nucleons and pions are built out of still more basic elementary particles, *quarks*. Nevertheless, the pion theory of the N–N interaction remains valid as far as it goes.

§9 PARTICLES AND ANTIPARTICLES

In special relativity the relation between the total energy E and the momentum p of a particle of rest mass m_0 is

$$E^2 = c^2p^2 + m_0^2c^4 \tag{9.1}$$

from which

$$E = \pm(c^2p^2 + m_0^2c^4)^{1/2}. \tag{9.2}$$

The question now arises as to the meaning of the negative-energy solutions.

Actually, it is clear from equation (9.2) that we must have

$$E > m_0c^2 \tag{9.3a}$$

or

$$E < -m_0c^2 \tag{9.3b}$$

i.e. the energy interval

$$-m_0c^2 < E < m_0c^2 \tag{9.4}$$

is forbidden. Thus a particle in the normal situation of positive energy would have to make a discontinuous transition if it were to drop down to a state of negative energy. Now such discontinuous transitions are forbidden in classical physics, so that if we postulate that at some instant in the past all particles were in positive-energy states then they must forever remain so. Thus the negative-energy states do not pose any basic difficulty to classical physics: they can simply be discarded.

Matters are less simple in quantum mechanics, where discontinuous transitions occur as a matter of course. It seems, in fact, that there is nothing to prevent a particle in a normal positive-energy state from jumping across the forbidden band of $2m_0c^2$ down to a state of negative energy. This was the problem faced by Dirac in 1928 when he tried to develop a quantum-mechanical theory of the electron that was consistent with special relativity. Dirac's solution was to invoke the Pauli exclusion principle and suppose that normally all negative-energy states were already filled with electrons, so that electrons in positive-energy states would be unable to cascade down. Since these negative-energy electrons would be uniformly distributed throughout space they would not, it was argued, give rise to any observable effect.

On the other hand, if an energy of at least $2m_ec^2$ could be communicated in some way to an electron in a negative-energy state then it would be raised to a positive-energy state and thus presumably become visible in the usual way. At the same time, the hole created in the sea of negative-energy states, consisting of the absence of a negative charge in an otherwise uniform and unobservable distribution, would manifest itself as a positively charged particle, later known as the *positron*†, e^+. Furthermore, if the state from which the electron was removed had momentum p and energy $-E$, the positron, or hole, would appear to have momentum $-p$ and energy $+E$, since the sea of negative-energy states will have lost momentum p and gained energy E. It then follows from equation (9.1) that we should attribute to the positron a mass m_e equal to that of the electron.

This simultaneous creation of an electron and a positron can be realised by γ rays whose energy is in excess of $2m_ec^2$ ($=1.022$ MeV)

$$\gamma \rightarrow e^+ + e^- . \tag{9.5}$$

Actually, although the electron and the positron materialise, as it were, out of the vacuum, the process cannot take place in empty space: the simultaneous conservation of momentum and energy requires that the γ photon be in interaction with some other particle, e.g. a nucleus (see Problem I.13).

Such so-called *pair production* occurs copiously when cosmic rays interact with matter at the surface of the earth, and it was in this way that the positron was discovered in 1932 by C D Anderson, in striking confirmation of Dirac's theory. Later, in 1934, it was found that some artificially produced isotopes could undergo β decays with the emission of positrons rather than electrons. (It is now known that a naturally occurring isotope, ^{40}K, β decays with the emission of positrons.)

Clearly, since a positron constitutes a hole in the sea of negative-energy electron states an encounter between a positron and an electron should lead to their mutual *annihilation*: the electron simply falls into the hole. The surplus energy of at least $2m_ec^2$ will be released in the form of *two* γ rays, it being impossible to conserve energy and momentum simultaneously with only one γ

$$e^+ + e^- \rightarrow 2\gamma . \tag{9.6}$$

In the foregoing discussion of negative-energy states we have been talking explicitly of the case of electrons, but it is obvious that everything we have said applies equally well to all fermions. Thus to a proton, p, there should correspond a negatively charged particle of equal mass, known as the *antiproton*, $\bar{p}$, while to a neutron, n, there should correspond another neutral particle having the same mass, known as the *antineutron*, $\bar{n}$. The existence of these particles was established, to no-one's surprise, in the 1950s, as soon as accelerators of sufficient energy became available, at least 2 GeV being

†To emphasise the difference between the electron and the positron, the former is sometimes referred to as a *negatron*, and the term 'electron' applied collectively to both particles.

required in the CM frame of reference. In §11 we shall encounter the neutrino and the antineutrino as another such pair.

A nucleon–antinucleon pair annihilates itself mainly through the creation of pions

$$p + \bar{p} \to \pi^+ + \pi^- \quad (9.7a)$$
$$p + \bar{p} \to 2\pi_0. \quad (9.7b)$$

The positron, antiproton and antineutron are referred to as the *antiparticles* of the corresponding particles, the electron, the proton and the neutron, respectively. One could in fact imagine them building up 'antiatoms' in which all the constituents are antiparticles of one or other of the constituents of ordinary atoms; we arrive thus at the concept of *antimatter*. The decision as to which to call particle and which to call antiparticle is made purely on the basis of which is the more frequently occurring in the vicinity of the earth: throughout our galaxy there is a strong preponderance of electrons, protons and neutrons.

For a long time it was speculated that there may be galaxies in which antimatter dominated: on the general grounds of symmetry arguments it was believed that there should be equal amounts of matter and antimatter in the universe. However, there is so far no evidence for this and it now seems in fact that there is a vast excess of matter over antimatter throughout the universe; the implications of this for cosmology and elementary particle physics are profound†.

It must be realised that this notion of antiparticles as being holes in a sea of particles having negative energy can apply only to fermions. In the case of bosons the Pauli principle does not operate, so there is no limit on the number of particles that can occupy any state, including those of negative energy. Thus in the case of bosons we have still not elucidated the meaning of the negative-energy states permitted by equation (9.2).

Dirac's hole theory is in fact out-dated and has been superseded by quantum field theory, in which the meaning of the negative-energy states is resolved quite naturally both for fermions and bosons. Nevertheless, the hole theory of fermions is still useful in that it accounts quite simply for the experimental fact that a fermion can never be created (or annihilated) without the simultaneous creation (or annihilation) of an antifermion. Bosons, on the other hand, can be created singly, e.g. in high-energy p–p scattering we can have

$$p + p \to p + n + \pi^+. \quad (9.8)$$

†See the article by F Wilczek in *Scientific American*, December 1980.

§10 RADIOACTIVITY

The phenomenon of radioactivity, the spontaneous emission of ionising radiations by certain nuclei, was mentioned briefly in §1. One type of radioactivity, γ decay, involves the transition of a nucleus in an excited state to a state of lower excitation of the same nucleus (possibly the ground state), the surplus energy being carried off by a photon (see §5). All other forms of radioactivity involve the spontaneous transformation of one nucleus into other nuclei of greater total binding energy, i.e. lower internal energy (see §5). In all these transmutations the total number of nucleons remains constant, but two main categories may be discerned.

(*a*) β instability. Here the unstable nucleus simply undergoes a spontaneous change in its charge, i.e. there is a transformation of a neutron into a proton or vice versa by virtue of the emission of an electron or a positron (known historically as *β rays* in this context). The mechanism responsible for this process is discussed in the next section.

(*b*) Fragmentation. Here the unstable nucleus breaks up into two nuclei, the sum of whose binding energies is greater than that of the original nucleus. The total number of neutrons and the total number of protons each remains unchanged in this process, unlike the case of β transformations. By far the most common disintegration of this type is α decay, characterised by one of the two fragments being a ^4He nucleus. It is also possible for an unstable nucleus to emit individual protons or neutrons, although this never happens for naturally occurring nuclides. Heavy nuclei can also break up into two nuclei of more or less comparable size, a process which is referred to as *spontaneous fission*; α decay is really a special case of this.

The energetics of both of these classes of transmutation will be discussed in Chapter III, with particular attention being paid to the conditions for stability or instability of a given nucleus. A transmutation will always be accompanied by a net loss of internal energy, which must be carried off by the particles formed in the process.

Soon after the discovery of radioactivity it was realised that the energy released by the decay of the radioactive elements occurring naturally in the earth would have a significant effect on its heat balance. More recently radioactive sources have served as small energy generators for use in space exploration, but for large-scale commercial generation of electricity they cannot possibly compete with fission reactors (Chapter V).

Decay laws. None of these radioactive processes (we are including here both γ decay and spontaneous fission) are instantaneous. It was known from the earliest studies of Rutherford and Soddy at the turn of the century that a finite timescale is involved, with the number of unstable nuclei remaining undecayed in a given sample decreasing exponentially with time, thus

$$n = n_0 e^{-\lambda t} \tag{10.1}$$

where n_0 is the number of unstable nuclei present at the arbitrary time $t = 0$ and λ is known as the *decay constant*, a quantity characteristic of the unstable nuclide in question, for the specified state of excitation (the nucleus *has* to be in an excited state for γ decay to occur, but other kinds of decay as well can take place from excited states (see the discussion on branched decays below)).

From the *decay law* (10.1), we find for the *activity* of a sample, i.e. the number of decays per unit time

$$\begin{aligned} R \equiv -\frac{dn}{dt} &= \lambda n_0 \, e^{-\lambda t} \\ &= \lambda n(t). \end{aligned} \tag{10.2}$$

This means that the fraction of the unstable nuclei present which decay in unit time interval, being equal to λ, is constant, independent of the time, and the same for all samples of the nuclide in question (we are supposing, of course, that only one species of unstable nucleus is present). Thus we are forced to adopt a probabilistic interpretation: λdt represents the probability that a given nucleus will decay in time dt. Confirmation of the essentially probabilistic nature of the decay process comes from the fact that the decay law (10.1) is only a statistical law, representing only average behaviour, and that significant fluctuations are observed.

We recall at this point that probability concepts are just as applicable to a population of living organisms as to a sample of decaying nuclei: life insurance would otherwise be impossible. Nevertheless, there are some very important differences. The probability of decay of a given nucleus in a given time interval is always the same, regardless of whether the nucleus was formed just a few minutes earlier in an accelerator experiment, or millions of years ago in some distant star. On the other hand, in a population of living systems the 'decay constant', i.e. the probability of death of a given individual in a given time interval, will increase with time, and the decline of the population will in fact be faster than exponential†. We may say, in effect, that nuclei do not age, their properties remaining unchanged right up to the instant of decay.

†In order to made the conditions as close as possible to those prevailing in our sample of decaying nuclei, we ignore the possibility of new members being added to the population through birth.

The decay of a nucleus is thus a truly random process, impossible to predict on the basis of the history of the nucleus. For example, of two identical unstable nuclei, one may decay in the next millisecond whereas the other may survive for millions of years before decaying, but we have no means of knowing this beforehand. In the case of living systems, on the other hand, it is possible, of course, to make meaningful qualitative predictions concerning the life expectancy of an individual, especially if we are in possession of certain information. The indeterminacy that prevails in the case of the unstable nucleus is completely alien to classical physics, although it arises quite naturally in the context of quantum mechanics. Similar remarks apply to the de-excitation of an excited atom by photon emission.

Now although it is impossible to predict how much longer a given nucleus will live, we can still define a *mean lifetime* $\bar{t}$. With n_0 unstable nuclei present at time $t=0$, let $\nu(t)\,\mathrm{d}t$ be the number that disintegrate in the interval of time t to $t+\mathrm{d}t$. Since this means that of the n_0 nuclei initially present, $\nu(t)\,\mathrm{d}t$ will have a lifetime between t and $t+\mathrm{d}t$, we see that the mean lifetime is given by

$$n_0\bar{t} = \int_0^\infty t\nu(t)\,\mathrm{d}t. \tag{10.3}$$

Now according to (10.2) we have

$$\nu(t)\,\mathrm{d}t = \lambda n(t)\,\mathrm{d}t \tag{10.4}$$

where $n(t)$ is the number of nuclei still surviving at time t. Then from (10.1)

$$\nu(t)\,\mathrm{d}t = \lambda n_0\,\mathrm{e}^{-\lambda t}\,\mathrm{d}t \tag{10.5}$$

from which

$$\bar{t} = \frac{1}{\lambda}. \tag{10.6}$$

The decay constant thus has the added significance of being the inverse of the mean lifetime.

It is important to elaborate somewhat on the concept of mean lifetime. If a particular unstable nucleus is present at a given instant, then the mean lifetime tells us how much longer we can statistically expect the nucleus to survive. Consider, for example, a reaction experiment in which we form a sample of radioactive nuclei of mean lifetime 10 years. This means that we can expect these nuclei to survive 10 years on the average. But then those nuclei that have still not decayed 30 years later can still be expected to survive another 10 years on the average. This is a consequence of the random nature of the decay process and stands in marked contrast to the prognostications that one would make in the case of living organisms.

Related to this concept of mean lifetime is that of *half-life*, the time that must elapse for the activity to fall to one half of its present value. From

(10.1) we see that this is given by

$$t_{1/2} = \frac{\ln 2}{\lambda} = \frac{0.693}{\lambda}$$
$$= 0.694\bar{t}. \tag{10.7}$$

The fact that the activity of a sample is always halved in the same time, regardless of its age, is a mathematical consequence, of course, of the exponential law (10.1), but basically it reflects the random nature of the decay process.

Branched decays. It often happens that more than one decay process is possible for a given unstable nucleus in a given state. This is particularly true in the case of nuclei that are in excited states, for then any β or fragmentation radioactivity will almost certainly be in competition with γ decay towards states of lower excitation. For each process we may define a *partial decay constant* λ_1, λ_2, etc, these representing the probability per unit time that a given nucleus will decay by the corresponding process†. The total probability of decay per unit time regardless of the particular process will be

$$\lambda = \lambda_1 + \lambda_2 + \ldots \tag{10.8}$$

and this total decay constant will determine the overall depopulation of the sample in the usual way according to (10.1). The mean lifetime of the nucleus will thus be

$$\bar{t} = \frac{1}{\lambda_1 + \lambda_2 + \ldots} \tag{10.9}$$

and likewise for the half-life. The fraction of decays that go through process i, for example, is λ_i/λ, a quantity which is independent of time.

Sequential decay. The daughter nucleus resulting from the decay of an unstable nucleus may itself be unstable, thus giving rise to a second decay, not necessarily of the same kind as the first. Indeed, it is possible to have a whole chain of successive decays, and in fact all radioactive nuclides heavier than lead belong to one or other such chain.

Let us consider the implications of the decay law (10.2) for such a chain of sequential decays; for simplicity we suppose that there are just two decays in the chain, with nucleus 1 decaying to nucleus 2, which in turn

†Note also that from a given initial state the same type of radioactivity could lead to different states of the same final nucleus. Each such transition is here regarded as a distinct process.

decays to stable nucleus 3. Then in an obvious notation we have

$$\frac{dn_1}{dt} = -\lambda_1 n_1 \tag{10.10a}$$

$$\frac{dn_2}{dt} = -\lambda_2 n_2 + \lambda_1 n_1. \tag{10.10b}$$

Solving this coupled system, the first nuclide is found to depopulate in the usual way

$$n_1(t) = n_{10}\,e^{-\lambda_1 t} \tag{10.11a}$$

where n_{10} denotes the number of nuclei of type 1 present at the time $t = 0$. Then from (10.10b) we get for the abundance of nuclide 2

$$n_2(t) = \frac{n_{10}\lambda_1}{\lambda_2 - \lambda_1}(e^{-\lambda_1 t} - e^{-\lambda_2 t}) + n_{20}\,e^{-\lambda_2 t} \tag{10.11b}$$

where n_{20} is the number of nuclei of type 2 present at the time $t = 0$. If we assume that this is zero, i.e. if we start counting at the instant of formation of the sample of nuclide 1, before any decays have taken place, then the second term in (10.11b) will be missing.

Assuming now that the origin of time corresponds to $n_{20} = 0$, we see that if nuclide 2 decays much faster than nuclide 1, i.e. if $\lambda_2 \gg \lambda_1$, then after a time $t_2 \gg 1/\lambda_2$ we shall have for (10.11b)

$$n_2(t) \simeq \frac{n_{10}\lambda_1}{\lambda_2}\,e^{-\lambda_1 t} \tag{10.12}$$

which means that nuclide 2 effectively decays at the same rate as its longer-lived precursor, nuclide 1. It is for this reason, of course, that many nuclides with lifetimes considerably shorter than the age of the earth are nevertheless found to occur naturally in the earth: they have long-lived precursors (see footnote p. 14). Referring to equations (10.12) and (10.11a) we see that the relative abundance of nuclides 1 and 2 becomes time independent for $t_2 \gg 1/\lambda_2$:

$$\frac{n_2(t)}{n_1(t)} = \frac{\lambda_1}{\lambda_2}. \tag{10.13}$$

The daughter nucleus 2 is effectively decaying as fast as it is formed, and a state of equilibrium has been established.

On the other hand, if nuclide 2 decays much more slowly than its precursor, i.e. $\lambda_2 \ll \lambda_1$, then (10.11b) becomes, with $n_{20} = 0$ again

$$n_2(t) \simeq n_{10}(e^{-\lambda_2 t} - e^{-\lambda_1 t}). \tag{10.14}$$

This means that the abundance of nuclide 2 at first increases according to

$(1 - e^{-\lambda_1 t})$ and then decreases according to $e^{-\lambda_2 t}$; effectively the two decays are decoupled in this situation.

§11 β RADIOACTIVITY: THE WEAK INTERACTION

By way of orientation table 11.1 lists a few of the several hundreds of known nuclear β decays. The symbol β^- and β^+ denote electron and positron emission, respectively, while EC refers to *electron capture*, a process that always competes with β^+ decay. Like this latter process electron capture involves the conversion of a proton into a neutron, doing so, however, not by the emission of a positron but by the capture of one of the atom's electrons.

The neutrino. One might have expected that the electrons emitted in a given β decay would be monoenergetic, the exact magnitude of their energy being determined by the energy difference between the initial and final nuclei. Even when the β decay leads to more than one state of the final nucleus, which is quite a common situation†, the β electron's energy spectrum might be expected to be at least discrete. Nevertheless, it is one of the remarkable features of β decay that the energy spectrum of the emitted

Table 11.1 Some typical β transitions.

Nucleus	Mode	T_0(MeV)	$t_{1/2}$	$\log_{10} ft$(s)
n	β^-	0.782	11.7 min	3.07
^{14}C	β^-	0.156	5692 yr	9.03
^{19}Ne	β^+ + EC	2.240	19.5 s	3.28
^{37}Ar	EC	−0.212	34.8 d	5.10
^{87}Rb	β^-	0.274	5×10^{10} yr	17.44
^{176}Lu	β^-	0.430	2.4×10^{10} yr	18.7
^{207}Tl	β^-	1.44	4.8 min	5.1
^{210}Bi	β^-	1.155	5 d	8.0

†None of the decays shown in table 11.1 have any significant branching.

electrons is continuous, with, however, a definite upper limit, the *end point*, which we denote by T_0 in table 11.1†. Typical β spectra are shown in figure 11.1.

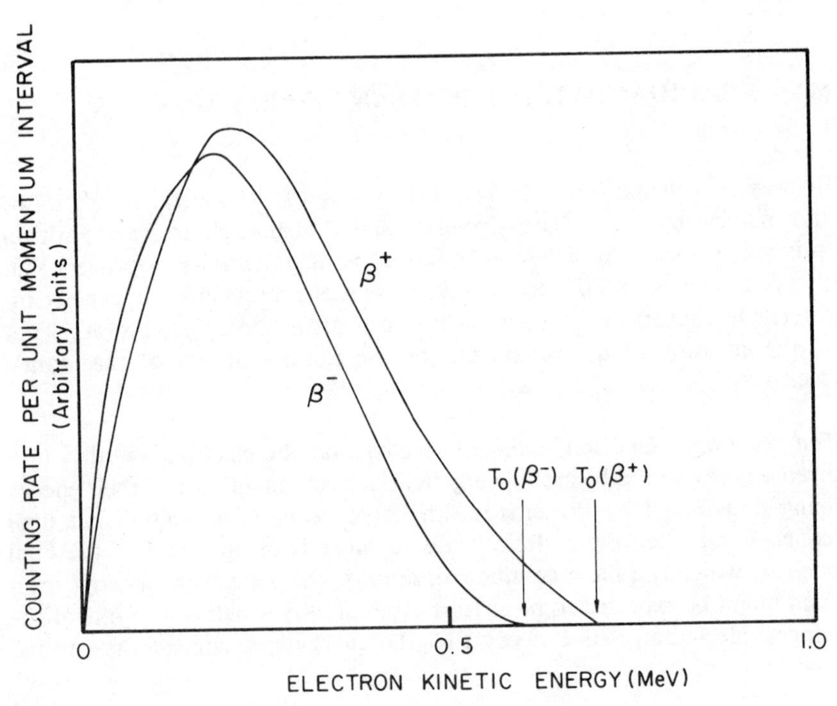

Figure 11.1 β^- and β^+ spectra of ^{64}Cu. This nucleus is exceptional in that it can undergo both types of β decay.

Pauli pointed out in 1930 that this continuous spectrum could still be reconciled with energy conservation if it was postulated that a second particle was emitted simultaneously with the electron, the total available energy being shared between the two in a continuously variable way. This particle, named the *neutrino*, would have to be neutral, and very much lighter than the electron‡. Furthermore, in order to explain the great difficulty encountered in detecting the neutrino, it had to be supposed that it interacted only very weakly with matter (only in 1956 was direct evidence for the

†Note that T_0 refers to *kinetic* energy, i.e. it does not contain the rest-mass energy of the electron. In the case of electron capture the meaning of T_0 changes (see equation (11.10)).
‡The neutrino associated with the β decay of the nucleus is not the same particle as the μ neutrino associated with the decay of the charged pion to the muon (p. 31).

neutrino's existence obtained; see p. 48). Finally, by postulating that the neutrino, like the electron, had spin $\frac{1}{2}\hbar$, it was possible to conserve angular momentum, as well as energy, in β decay (the old proton–electron view of the nucleus would require rather that the neutrino have integral or zero spin).

Since we have seen that the electron is not a constituent of the nucleus, we cannot regard β decay as some sort of leakage of electrons from the nucleus, in the manner of α decay. Rather, we must think of it as involving the simultaneous *creation* of an electron–neutrino pair, together with a corresponding change in the charge of a nucleon. The basic β^- decay process is then

$$\mathrm{n} \rightarrow \mathrm{p} + \mathrm{e}^- + \bar{\nu} \tag{11.1}$$

and the basic β^+ process

$$\mathrm{p} \rightarrow \mathrm{n} + \mathrm{e}^+ + \nu \tag{11.2}$$

where the symbols ν and $\bar{\nu}$ denote respectively 'neutrino' and 'antineutrino', each being the antiparticle of the other, in exactly the same way as the electron and positron are. (It should be noted that the β decay of the free proton is energetically impossible, so that the process (11.2) can only take place for protons bound in nuclei—see §16. On the other hand, we see from table 11.1 that the free neutron *is* β unstable.)

It might be asked how a neutral particle can be distinguished from its antiparticle, since both will be in the same charge state, unlike electrons and positrons. Indeed, there exists the theoretical possibility that the neutrino and the antineutrino are identical, but it has been shown experimentally that they can be distinguished (see p. 49). The physical difference between the two is now known to consist of the neutrino always being polarised with its spin opposite to the momentum vector, while the antineutrino always has its spin parallel to the momentum vector†. It should be realised that attributing the name 'antineutrino' to the particle emitted in the neutron decay rather than the proton decay is largely a matter of convention; the choice made is such that if we also regard the electron as a particle and the positron as an antiparticle, then the *difference* between the total number of particles and the total number of antiparticles is conserved in all nuclear β decay processes.

Fermi theory of weak interactions. Very shortly after this picture of β decay emerged (in 1934, to be precise) Fermi perceived an analogy with the emission of light by an atom. That is to say, just as a photon is created when an electron changes its orbital state in an atom, so an electron–neutrino pair is created when a nucleon changes its charge state, i.e. when a neutron is

†For further details on this and many other aspects of this section the reader is referred to the book by Frauenfelder and Henley (General Bibliography).

transformed into a proton or vice versa. Pursuing this analogy with electromagnetic transitions Fermi was able to construct a theory which remains substantially intact to this day, although it must be stressed that the analogy is only formal. In particular, we shall see that the basic force responsible for β decay cannot be electromagnetic in origin†.

We shall not be able to enter here into the details of the Fermi theory of β decay, but will limit ourselves to a rapid sketch of some of its main features. First, let us try to understand the bell shape of the typical β spectrum (figure 11.1). This will be determined by the way in which the total available energy Q, associated with the change in nuclear binding energy (§16), is shared between the neutrino and the electron.

We have

$$Q = E_e + E_\nu \tag{11.3}$$

where

$$E_e = (m_e^2c^4 + p_e^2c^2)^{1/2} \tag{11.4}$$

is the total relativistic electron energy, and

$$E_\nu = (m_\nu^2c^4 + p_\nu^2c^2)^{1/2} \tag{11.5}$$

is the total neutrino energy, m_ν being the neutrino mass. We shall have for the maximum total energy of the electron

$$E_e^{max} = Q - m_\nu c^2 \tag{11.6}$$

from which the maximum kinetic energy of the electron

$$\begin{aligned} T_0 &\equiv E_e^{max} - m_e c^2 \\ &= Q - (m_e + m_\nu)c^2. \end{aligned} \tag{11.7}$$

Likewise, the maximum total energy of the neutrino is

$$E_\nu^{max} = Q - m_e c^2 \tag{11.8a}$$

from which the maximum kinetic energy of the neutrino is

$$T_0^\nu = E_\nu^{max} - m_\nu c^2 = T_0. \tag{11.9}$$

In electron capture the emitted neutrino has a unique energy. With the electron being captured from an atomic orbital of energy ε (negative), we have

$$E_\nu = Q + m_e c^2 + \varepsilon. \tag{11.8b}$$

†In fact, the currently accepted theory of Weinberg and Salam indicates that the interaction responsible for β decay and the electromagnetic interaction have a common origin, although they remain phenomenologically quite distinct from the point of view of the nucleus.

Then, in this case, the neutrino's unique kinetic energy is

$$T_0^{\nu} = T_0 + 2m_e c^2 + \varepsilon. \tag{11.10}$$

This is how the T_0 of table 11.1 is to be interpreted in the case of electron capture, and we see how negative values are possible (see also §16).

To a large extent the distribution of Q between E_e and E_ν will be determined by purely statistical considerations, the phase space available to the electron being proportional to p_e^2, and likewise that for the neutrino to p_ν^2. Thus the spectrum shape will depend mainly on the factor

$$p_e^2 p_\nu^2 \propto p_e^2(E_\nu^2 - m_\nu^2 c^4) = p_e^2[(Q - E_e)^2 - m_\nu^2 c^4] \equiv X(E_e). \tag{11.11}$$

This goes to zero both at the low-energy end ($p_e = 0$) and the high-energy end ($E_e = E_e^{\max} = Q - m_\nu c^2$), from which we get the bell-shaped curve. However, it must be realised that the Coulomb interaction between the decay electron (or positron) and the nucleus modifies the spectrum shape considerably.

From equation (11.11) we have for the gradient of the statistical factor at the end point T_0 of the spectrum

$$\begin{aligned} \left(\frac{\mathrm{d}X(E_e)}{\mathrm{d}E_e}\right)_{E_e = E_e^{\max}} &= -2(p_e^{\max})^2(Q - E_e^{\max}) \\ &= -2m_\nu c^2 (p_e^{\max})^2. \end{aligned} \tag{11.12}$$

Thus careful measurement of the shape of the β spectrum at the end point will serve to determine the neutrino mass, m_ν. Until recently all measurements were consistent with $m_\nu = 0$, but in the last few years there has been some evidence for a value of 20 to 40 eV. A non-zero rest mass for the neutrino would have enormous cosmological implications, but at the time of writing (early 1985) the situation is far from clear. In any case, it will be sufficient for our purposes to set $m_\nu = 0$.

Let us now integrate the statistical factor

$$\begin{aligned} X(E_e) &= p_e^2(Q - E_e)^2 \\ &= \frac{1}{c^2}(E_e^2 - m_e^2 c^4)(Q - E_e)^2 \end{aligned} \tag{11.13}$$

over the entire spectrum. We shall have

$$\int_{m_e c^2}^{E_e^{\max}} X(E_e)\, \mathrm{d}E_e \simeq \frac{1}{30c^2} Q^5 \tag{11.14}$$

if $Q \gg m_e c^2$. Thus we see that the statistical factor implies that the decay rate increases very strongly with the available energy Q.

Actually, this analysis ignores the influence of the Coulomb force on the spectrum shape. If we take this into account in integrating over the spectrum shape the Q^5 factor of equation (11.14) is replaced by the so called

f factor. This, too, depends very strongly on Q, but cannot be expressed in any simple analytic form; it has been extensively tabulated as a function of Q and Z, for both electrons and positrons.

The product of this function f for a given β decay, and the corresponding half-life $t_{1/2}$, known as the ft value of the decay, should be independent of any energy-dependent statistical factors, and might therefore be expected to be the same for all decays. A glance at the last column of table 11.1 shows that this is far from the case. This variation of the ft value from one decay to another is a consequence of a dependence of the decay rate on the structure of the nuclei involved, and in particular on the similarity, or 'overlap', of the respective wavefunctions of the initial and final nuclei (see equation (12.30)). In this respect it is significant that the ft values are minimal, i.e. the decay rates are maximal (after correction for statistical factors), in the case of decays between mirror nuclei (see in table 11.1 the decays of the neutron and $^{19}_{10}$Ne).

We are now in a position to compare the strength of the interaction responsible for β decay with the electromagnetic interaction. Taking the most favourable case of the neutron decay, we compare its half-life, 11.7 minutes, with that of an electromagnetic transition with a similar energy release, e.g. a nuclear γ decay. Now these latter typically have half-lives of the order of 10^{-15} s, so it is clear that in β decay we are dealing with an interaction that is many times weaker than the electromagnetic interaction.

This interaction responsible for β decay is known as the *weak interaction*, and it is one of the four basic interactions occurring in nature, the other three being the nuclear (strong), electromagnetic and gravitational interactions†. All these four forces have a role to play in nuclei, although the last is significant only for the supergiant nuclei, i.e. neutron stars (see §15). Lest it be thought that β decay is a somewhat peripheral phenomenon from the point of view of the physics of the nucleus, however important it may be in terms of the study of the basic forces of nature, we would point out that it is of absolutely fundamental significance in determining some of the large-scale macroscopic manifestations of nuclear physics: we shall find that without the weak interaction the synthesis of the chemical elements in the stars (Chapter VI), and the control of nuclear reactors (Chapter V) would both be impossible.

We have seen that both the strong and electromagnetic interactions are mediated by the exchange of bosons (pions and photons, respectively). It

†Although weaker than the electromagnetic interaction the weak interaction is usually regarded as being stronger than the gravitational interaction. However, the comparison is a little difficult to make since we shall see in §15 how in large aggregates of matter the gravitational force can dominate the nuclear force, which is generally considered as the strongest member of the hierarchy.

is now known that the weak interaction arises likewise from the exchange of so-called intermediate bosons, the charged $W^{\pm}$ and the neutral Z^0.

We shall not write out explicitly the form of the weak-interaction Hamiltonian H_β responsible for β decay, but will simply point out that it is strongly non-invariant under the parity operation, i.e. equation (7.10) does not hold. That is why the neutrino and the antineutrino are polarised (the electron and the positron are also partially polarised).

Inverse processes. Since the creation of a particle is formally equivalent to the destruction of its antiparticle, we see, rewriting equation (11.2) appropriately, the possibility of the process

$$p + e^- \rightarrow n + \nu. \tag{11.15}$$

This is, in fact, the prototype of the electron capture process that we have already mentioned.

Various forms of electron capture can be envisaged, but the most common consists simply of capture from an atomic orbit, usually the K shell, but also, with diminishing probability, from the L, M, etc, shells. Because of the electron hole that is thereby created, electron capture is associated with the emission of x-rays characteristic of the atomic shell from which capture occurred.

Orbital-electron capture always competes with β^+ emission, but the dependence on the available energy is quite different, capture being much less important than β^+ emission for large T_0, but predominating for small T_0. It is for this latter reason that it is very difficult to see positron emission in long-lived nuclei, and for a long time it was associated exclusively with artificial isotopes. However, a very weak β^+ emission has since been found in the naturally occurring ^{40}K. Because the energy requirements for orbital-electron capture to occur are less stringent than for β^+ decay, essentially because an electron is being destroyed, rather than created, we can have capture without there being any β^+ emission at all, although the converse situation is not possible (see §16).

In fact, even if the nucleus in question were completely stable with respect to orbital-electron capture, the process (11.15) could still take place, in principle, provided the nucleus were bombarded with a beam of electrons having sufficient energy. Of course, the inverse decay would also occur, so eventually an equilibrium would be established between the nuclides represented by the two sides of equation (11.15), the relative concentrations depending on the beam intensity. Unfortunately, it has not been possible to detect process (11.15) in the laboratory against the background of all the other processes associated with the electron beam.

However, there exists an astrophysical context in which normally stable nuclei do capture electrons at a significant rate: the formation of a neutron star from the gravitational collapse of a white dwarf, as described in Appendix D. We shall examine this process in more detail in §16, and simply note for the moment that the essential feature of this situation is that the neutron-rich nucleus, once formed by electron capture, is unable to β decay back to the original nucleus because the electron gas in which it is immersed is degenerate.

Rewriting equation (11.1) in exactly the same way as we did for equation (11.2), we obtain

$$n + e^+ \rightarrow p + \bar{\nu} \qquad (11.16)$$

and see the possibility of 'positron capture'. In fact such a process is impossible to realise at a measurable rate under terrestrial conditions, but it is believed to take place in the interiors of very hot stars, positron–electron pairs being created by the substantial flux of black-body photons having an energy in excess of the necessary 1.02 MeV (the temperature corresponding to a kT energy of this magnitude is 10^{10} K).

Rather than transpose the electron or positron to the left-hand sides of equations (11.1) or (11.2), respectively, as we did in writing equations (11.16) and (11.15), we could transpose the antineutrino or the neutrino, and obtain

$$n + \nu \rightarrow p + e^- \qquad (11.17)$$

$$p + \bar{\nu} \rightarrow n + e^+ \qquad (11.18)$$

indicating thereby the possibility of inducing β transitions by neutrino fluxes of sufficient energy and intensity. Such neutrino fluxes could themselves be obtained from a β decaying source, but assuming a distinction between ν and $\bar{\nu}$ we see that the reaction (11.17) could only be induced by a β^+ emitter, and (11.18) by a β^- emitter.

The strongest available source of β^- emission, and hence of $\bar{\nu}$, is a nuclear reactor (the β^- decays occur in the fission products; see §23). Reines and Cowan (1956) showed that the reaction (11.18) did indeed take place in tanks of water placed close to a reactor, providing thereby the first direct evidence for the antineutrino.

With the existence of the antineutrino confirmed, there could be little doubt about the neutrino, but there was still considerable interest in realising the reaction (11.17), as we shall see. We address ourselves first to the problem that we cannot use a target of free neutrons, since these are unstable anyway. It transpired that the most convenient target to use in this case was the stable ^{37}Cl, the reaction to be looked for then being

$$^{37}\text{Cl} + \nu \rightarrow {}^{37}\text{Ar} + e^- . \qquad (11.19)$$

The ^{37}Ar is, of course, unstable, decaying back to ^{37}Cl through electron capture, but an equilibrium should be established, the concentration of ^{37}Ar in any sample of ^{37}Cl depending on the neutrino flux to which it is exposed. The technique developed by R Davis consists of taking ^{37}Cl in the form of large tanks of tetrachlorethylene (C_2Cl_4, dry-cleaning fluid) and looking for traces of radioactive ^{37}Ar therein. This arrangement was first used to show that ν and $\bar{\nu}$ are indeed different particles. If these were identical then close to a reactor the reaction (11.19) should proceed just as readily as did the reaction (11.18). Davis found no evidence for any reaction in ^{37}Cl.

Having established that ν and $\bar{\nu}$ are indeed distinct, the question arises as to a suitable source of ν by which reactions (11.17) or (11.19) might be induced. No terrestrial source is remotely strong enough to give rise to a detectable reaction rate, but it was estimated that the flux of neutrinos coming from the sun (originating in fusion reactions that give rise to nuclei that are proton-rich and hence β^+ unstable; see §27) might be sufficient. Since the flux of solar neutrinos, however weak or strong it may be, is all-pervading and cannot be switched off, it follows that any sample of ^{37}Cl must always contain a certain equilibrium concentration† of the unstable ^{37}Ar. Such evidence for solar neutrinos has now been found in the Davis experiment, but the apparent flux is significantly lower than what is predicted by most solar models.

We conclude this discussion by looking once more at the question of the distinguishability of the neutrino and antineutrino. If they were identical then the following sequence could be envisaged

$$\begin{array}{l} n \rightarrow p + e^- + \boxed{\begin{array}{c} \nu \\ + \\ n \end{array}} \rightarrow p + e^- \end{array}$$

i.e.

$$2n \rightarrow 2p + 2e^-. \tag{11.20}$$

Thus we should have the possibility of neutrino-less β decay with Z changing by *two* units at a time. Such double β decay has never been observed.

†One might expect slight variations in this equilibrium concentration over a six-month period as the earth–sun distance varies. On the other hand there should be no difference between day and night, the solar neutrinos passing right through the earth with negligible attenuation.

PROBLEMS FOR CHAPTER I

I.1 A beam of deuterons of non-relativisitic energy is elastically scattered by a hydrogen target. Show that according to classical mechanics the scattering angle cannot exceed 30° in the laboratory system. On the other hand, show that when a beam of protons is scattered by a deuterium target there is no limitation on the scattering angle.

I.2 Show that when a particle of non-relativistic energy is elastically scattered by another particle of the same mass the angle betweeen their final directions in the laboratory system is always 90°, according to classical mechanics.

I.3 A beam of 6 MeV α particles is incident on a thin gold target, 2 mg cm^{-2}.

(*a*) Calculate the distance of closest approach of an α particle to a gold nucleus, and compare with the radius of the latter ($A = 197$).

(*b*) Calculate the de Broglie wavelength of relative motion, and compare with the radius of the gold nucleus.

(*c*) Calculate the impact parameter for a scattering angle of 60° in the centre of mass (CM system). What is the corresponding angular momentum in units of $\hbar$?

(*d*) What fraction of particles will be scattered through more than 60° (CM system)?

(*e*) What is the probability that a given α particle will be scattered twice, each time through an angle greater than 10° (CM system)?

I.4 Fill in the gaps between equation (2.8) and equation (2.9).

I.5 Referring to figure 2.2 and making use of equations (2.4) and (2.9), deduce a general expression for the minimum distance of approach OA in the case of a projectile of impact parameter b. Derive equation (2.18) as a limiting case.

I.6 If the proton–electron model of the nucleus were valid, show that all neutral atoms would contain an even number of fermions.

I.7 The RMS radius, η, of the charge distribution $\rho(r)$ of a spherical nucleus is defined by

$$\eta^2 = \frac{\int r^2 \rho(r)\, \mathrm{d}^3 r}{\int \rho(r)\, \mathrm{d}^3 r}$$

where both integrals go over the entire nuclear volume.

(*a*) Show that for a uniform distribution, radius R,

$$\rho = \begin{cases} \rho_0 & r < R \\ 0 & r > R \end{cases}$$

we have

$$\eta = \sqrt{\frac{3}{5}}\, R.$$

(*b*) Show that for a trapezoidal distribution,

$$\rho = \begin{cases} \rho_0 & r < R - \frac{t}{2} \\ \rho_0\left(\frac{1}{2} + \frac{R-r}{t}\right) & R - \frac{t}{2} < r < R + \frac{t}{2} \\ 0 & r > R + \frac{t}{2} \end{cases}$$

we have, for $t \ll R$

$$\eta = \sqrt{\frac{3}{5}}\, R\left[1 + \frac{7}{24}\left(\frac{t}{R}\right)^2 + \mathrm{O}\left(\frac{t}{R}\right)^3\right].$$

In both cases express ρ_0 in terms of Ze and the parameters of the distribution.

I.8 Show that the overall internal energy of a neutral atom is given by

$$E_{\mathrm{at}} = (M_{\mathrm{at}} - NM_{\mathrm{n}} - ZM_{\mathrm{H}})c^2 + ZE_{\mathrm{at}}(H)$$

where $E_{\mathrm{at}}(H)$ is the internal energy of the hydrogen atom.

I.9 Calculate the binding energy of the electrons of a uranium atom.

I.10 Even–even nuclei (i.e. N and Z both even) in their ground state always have $J = 0$ for the total angular momentum, and positive parity. Show that the residual nucleus after α decay can only be in a state $J^\pi = 0^+, 1^-, 2^+, 3^-$, etc, and never $0^-, 1^+, 2^-, 3^+$, etc. (The α particle is always in its ground state.)

I.11 For a two-nucleon system we define the following quantities:

$$\boldsymbol{r} = \boldsymbol{r}_1 - \boldsymbol{r}_2$$
$$\boldsymbol{p} = \tfrac{1}{2}(\boldsymbol{p}_1 - \boldsymbol{p}_2)$$
$$\boldsymbol{l} = \boldsymbol{r} \times \boldsymbol{p}$$
$$\boldsymbol{S} = \boldsymbol{s}_1 + \boldsymbol{s}_2.$$

Consider the following scalar quantities on which the interaction potential could conceivably depend

$$\boldsymbol{r}\cdot\boldsymbol{p},\ \boldsymbol{r}\cdot\boldsymbol{S},\ \boldsymbol{r}\cdot\boldsymbol{l},\ \boldsymbol{l}\cdot\boldsymbol{S},\ \boldsymbol{l}\cdot\boldsymbol{p},\ \boldsymbol{p}\cdot\boldsymbol{S}.$$

Which of these terms remains invariant under the parity operation? (Note that the first three of these change sign under the operation of time reversal and therefore are usually discarded for this reason.)

I.12 Let the n–p interaction (supposed to be purely central) be represented by a square well of radius $b = 2$ fm. Calculate the depth V_0 of this well for triplet and singlet S states, respectively, given that (*a*) the binding energy of the deuteron is 2.22 MeV and (*b*) in the limit of vanishing bombarding energy, the singlet S phaseshift in n–p elastic scattering behaves as

$$\lim_{k \to 0} \frac{\delta(^1S)}{k} = 23 \text{ fm}$$

where k is the wavenumber of relative motion.

I.13 Show that the pair production process

$$\gamma \to e^+ + e^-$$

is impossible in the absence of any other particle.

I.14 Consider the presence of a muon (§8) in the ground-state Bohr orbit of a lead atom ($Z = 82$). Using the expression (2.20), with $m_\mu = 207\ m_e$ and $n = 1$, show that the radius of this orbit is less than that of the nucleus. (Quantum mechanically this means that the muon spends a considerable fraction of its short lifetime inside the nucleus before decaying, so that there will be a significant modification of the expression (2.19) for the energy of the orbit, this expression being based on the assumption of a point nucleus.)

CHAPTER II

BINDING ENERGY OF THE NUCLEUS

In this chapter we discuss some of the attempts that have been made to calculate the measured binding energies of nuclei, and in particular to account for the main trends observed in their variation from one nucleus to another.

§12 INDEPENDENT-PARTICLE MODEL

The concept of the binding energy B of a nucleus was introduced in §5, where we showed that it is related to the internal energy E by

$$B = -E \tag{12.1}$$

with E being given as the eigenvalue of the Schrödinger equation (5.2) of the nucleus

$$H\Psi(r_1, r_2, \ldots, r_A) = E\Psi(r_1, r_2, \ldots, r_A). \tag{12.2}$$

We recall that the Hamiltonian H of the nucleus is given by

$$H = \hat{T} + \hat{V} \tag{12.3}$$

where

$$\hat{T} = -\frac{\hbar^2}{2M}\sum_{i=1}^{A} \nabla_i^2 \tag{12.4}$$

is the kinetic energy operator, and

$$\hat{V} = \sum_{i>j}^{A} \sum_{j=1}^{A} V_{ij} \tag{12.5}$$

is the potential energy operator. It was stated that if the Schrödinger equation could be solved, not only would we have the energy E but the wavefunction Ψ would give us all other properties as well.

However, we recall from §5 that solving this equation is impossible at the present time for all nuclei with more than three or four nucleons. It is not surprising, therefore, that rather than starting from the basic Schrödinger equation, a somewhat less ambitious programme has been followed, particularly in the past, and that most attempts to understand nuclear structure have been made in terms of various phenomenological models.

One model in particular should be mentioned: the *shell model* or *independent-particle model*, the basic idea of which is that the motion of any one nucleon, the ith, say, does not depend on the detailed motion of any of the other nucleons. Rather, the effect of all the other nucleons is just to generate a mean potential field $V(\boldsymbol{r}_i, q_i)$, where q_i denotes coordinates of the ith nucleon other than position, e.g. spin and momentum. Thus the force on the ith nucleon is simply a function of its own coordinates and does not depend on the instantaneous configuration of the other nucleons: we say that *correlations* between nucleons are ignored.

The expression (12.5) for the potential energy then becomes replaced by

$$\hat{V}_0 = \sum_{i=1}^{A} V(\boldsymbol{r}_i, q_i) \tag{12.6}$$

so that the exact Hamiltonian (12.3) is replaced by the model Hamiltonian

$$H_0 = \sum_{i=1}^{A} h_i \tag{12.7}$$

where

$$h_i = -\frac{\hbar^2}{2M} \nabla_i^2 + V(\boldsymbol{r}_i, q_i). \tag{12.8}$$

Instead, then, of solving the exact Schrödinger equation (12.2) we solve the model Schrödinger equation

$$H_0 \Psi_0 = E_0 \Psi_0. \tag{12.9}$$

The essential feature of H_0, as given by equations (12.7) and (12.8), is that each term h_i depends only on the coordinates of a single nucleon. It then follows that the model wavefunction Ψ_0 factorises into A functions $\phi_i(\boldsymbol{r}_i, q_i)$, each of which is associated with just one nucleon:

$$\Psi_0(\boldsymbol{r}_1, \ldots, \boldsymbol{r}_A) = \prod_{i=1}^{A} \phi_i(\boldsymbol{r}_i, q_i) \tag{12.10}$$

(we are ignoring here the effect of the Pauli principle, which complicates matters somewhat; see Problem (II.1)). To see this, we substitute equations (12.7) and (12.10) into equation (12.9). The left-hand side can be written as

$$\left(\sum_{i=1}^{A} h_i\right) \prod_{i=1}^{A} \phi_i = (h_1\phi_1) \prod_{i=2}^{A} \phi_i + \left(\sum_{i=2}^{A} h_i\right) \prod_{i=1}^{A} \phi_i$$

$$= \sum_{i=1}^{A} (h_i\phi_i) \frac{1}{\phi_i} \Psi_0 = \sum_{i=1}^{A} \left(\frac{1}{\phi_i} h_i\phi_i\right) \Psi_0. \qquad (12.11)$$

Thus equation (12.9) reduces to

$$\sum_{i=1}^{A} \frac{1}{\phi_i} h_i\phi_i = E_0. \qquad (12.12)$$

Since no two terms on the left-hand side here can depend on the same variable, it follows that each term must in fact be a constant, since their sum is. Thus we can write for each nucleon

$$h_i\phi_i = \varepsilon_i\phi_i \qquad (12.13)$$

where ε_i is a constant, with

$$E_0 = \sum_{i=1}^{A} \varepsilon_i. \qquad (12.14)$$

Thus we have proven that the model Schrödinger equation (12.9) does indeed have a solution of the separated form (12.10), provided each ϕ_i satisfies an equation of the form (12.13). But this latter equation has just the form of a Schrödinger equation for a single nucleon i, the Hamiltonian of which is h_i. It follows that the associated constant ε_i can be interpreted as the energy of the nucleon i: it is referred to as the *single-particle energy*, while the function $\phi_i(\boldsymbol{r}_i, q_i)$ is known as the *single-particle wavefunction.*

If the so-called single-particle potential $V(\boldsymbol{r}_i, q_i)$ is postulated by the model (or else derived from more basic principles), the solution of the single-particle equation (12.13) presents no difficulty at all. This is because only one particle is involved at a time (although there is, of course, one such equation for each particle), unlike the exact Schrödinger equation (12.2), where all A nucleons are coupled together.

Now although such a model is easily soluble, there does not seem to be any reason why it should constitute a valid description of the nucleus, for it does not appear to be possible that the motion of one nucleon should be independent of the detailed motion of the other nucleons, particularly in view of the fact that nucleons can interact only with their nearest neighbours. This last observation is a consequence of the simple fact discussed in §4 that all nuclei have more or less the same density. With the parameter r_0 appearing in equation (4.2) taking the approximate value of

1.15 fm, it follows that the mean distance between nucleons in a nucleus will be about 2.3 fm, which is of the same order as the range of the nucleon–nucleon interaction.

In the case of the electrons of an atom the existence of such a common potential field is easily understood. A large part of the field comes from the Coulomb potential $-Ze^2/r$ of the nuclear charge, while the electron–electron interactions can be partially smoothed out into an additional common field, the long range of the Coulomb force ensuring that each electron interacts with many others. But there is nothing inside the nucleus that plays the same role there as does the charge of the nucleus inside the atom, and it would seem impossible for the nucleon–nucleon forces to smooth themselves out into a common field if each nucleon interacts only with its nearest neighbours. Another, and as we shall see, more instructive, way to look at the paradox is to say that one would expect the strong short-range force to lead to a rapid sharing of energy and momentum between colliding nucleons.

Nevertheless, nuclei do display some of the characteristic properties of the independent-particle model, similar to those displayed by atoms. To understand these properties, consider the case where the single-particle potential has the spherical-oscillator form

$$V(r) = \tfrac{1}{2}M\omega^2 r^2. \tag{12.15}$$

Solving the corresponding single-particle equation (12.13) in spherical coordinates leads to solutions of the form

$$\phi_{nlm}(\boldsymbol{r}) = A_{nl}\rho^l L_n^{l+\frac{1}{2}}(\rho^2)\exp\left(-\frac{\rho^2}{2}\right) Y_l^m(\theta, \phi) \tag{12.16}$$

where

$$\rho = \sqrt{\frac{M\omega}{\hbar}}\, r \tag{12.17}$$

$L_n^{l+\frac{1}{2}}$ is a Laguerre polynomial, n a positive integer or zero, (l,m) the usual quantum numbers of orbital angular momentum and A_{nl} a normalisation constant.

The corresponding energy eigenvalues are given by

$$\varepsilon_{nl} = (2n + l + \tfrac{3}{2})\hbar\omega \tag{12.18}$$

which may be conveniently rewritten as

$$\varepsilon_{\mathcal{N}} = (\mathcal{N} + \tfrac{3}{2})\hbar\omega \tag{12.19}$$

where

$$\mathcal{N} = 2n + l. \tag{12.20}$$

The different possible values of this, 0, 1, 2, 3, etc, define the different energy levels, which will be uniformly spaced with a separation of $\hbar\omega$. In general we see that different l values will be associated with different $\mathcal{N}$ values, and since each l value has associated with it $(2l+1)$ different m quantum numbers it follows that the different levels will have different degeneracies. Bearing in mind that for each set of (n, l, m) quantum numbers there can be two spin states for a given type of fermion, the overall degeneracy of the different levels will be as indicated by the column labelled d in table 12.1.

Table 12.1 Level scheme of spherical harmonic oscillator.

$\mathcal{N}$	l	d		ν
0	0	2		2
1	1	6		8
2	0 2	2 + 10	12	20
3	1 3	6 + 14	20	40
4	0 2 4	2 + 10 + 18	30	70
5	1 3 5	6 + 14 + 22	42	112

For simplicity let us suppose that this oscillator model is being applied to the electrons of an atom, so that only one kind of fermion is involved. We can picture the atom as being built up by feeding electrons into the oscillator well, which is imagined to be predetermined, i.e. it is not perturbed by the addition of electrons. Since we are interested only in the ground state, electrons will always go into the lowest available single-particle state, consistent with the Pauli principle. Thus the first two electrons will go into the $\mathcal{N}=0$ level, the next six into the $\mathcal{N}=1$ level, and so on.

The quantity labelled ν in the last column of table 12.1 represents the total number of particles present when the corresponding level is completely filled; in other words, it denotes the maximum number of particles that can be accommodated while leaving all higher-lying levels completely unoccupied. It follows that atoms whose number of electrons is one of the indicated values of ν will have an exceptional stability, since additional electrons will be required to start filling the next higher-lying level, with a consequent loss of binding energy.

This existence of particle numbers for which the system in question has an exceptional stability is one of the hallmarks of independent-particle structure, and it is, of course, manifested by atoms, the noble gases constituting the exceptionally stable atoms (actually, the independent-particle structure of atoms is displayed not only by the existence of the noble gases but by the whole periodic table). However, the atomic numbers of the noble gases do not correspond to the ν of table 12.1, but are rather given by 2, 10, 18, 36, 54 and 86. The difference arises from the fact that the common field in which the electrons of an atom move, being the resultant of the Coulomb field of the nucleus and the smoothed effect of the electron–electron interactions, does not have an oscillator form. The result is a quite different ordering of the levels, and in fact the exact degeneracies of different l values that we have found with the oscillator model no longer prevail. Nevertheless, the different states still cluster together into distinct groups, known as *shells*, and our qualitative arguments still hold.

By 1948 it had become apparent that nuclei, too, exhibited independent-particle features, with exceptional stability being associated with the following numbers of neutrons and protons, respectively

$$N = 2,\ 8,\ 20,\ 28,\ 50,\ 82,\ 126$$
$$Z = 2,\ 8,\ 20,\ 28,\ 50,\ 82.$$

A problem arose, however, in connection with the form of the single-particle potential $V(r_i, q_i)$ that would give rise to these so-called *magic numbers*. The oscillator potential would clearly not do, but then we do not expect this form for the field generated by the nucleons. Rather, in view of the short range of the nucleon–nucleon force, the potential should follow more or less the matter distribution, and it could thus be conveniently represented by

$$V_{SW}(r) = \frac{-V_0}{1 + \exp[(r - R)/a]} \tag{12.21}$$

which is just the Fermi function of equation (4.1), although it is referred to as the *Saxon–Woods* potential. To this should be added a Coulomb term V_C in this case of protons (see Problem II.2). However, realistic potentials of this form do not yield the required magic numbers, and in fact it was found to be impossible to reproduce them with any form of potential that is a function solely of r. (We are tacitly supposing spherical symmetry; see below.)

The key to the problem was provided by including in the single-particle potential a so-called *spin–orbit* term, i.e. a coupling between the spin and the orbital angular momentum of the nucleon in question. One writes

$$V(r_i, q_i) = V_{\text{SW}}(r_i) + V_{\text{C}}(r_i) + V_{\text{so}}(r_i)s_i \cdot l_i. \tag{12.22}$$

The effect of this is to remove the degeneracy between single-particle states with total angular-momentum quantum number $j = l + \frac{1}{2}$, and states with $j = l - \frac{1}{2}$. To see this we denote the single-particle states by ϕ_{njlm}, and note that

$$\hat{j}^2 \phi_{njlm} = j(j+1)\hbar^2 \phi_{njlm} \tag{12.23}$$

where

$$\hat{j} = \hat{l} + \hat{s} \tag{12.24}$$

so that

$$\hat{j}^2 = (\hat{l} + \hat{s})^2 = \hat{l}^2 + \hat{s}^2 + 2\hat{l} \cdot \hat{s}. \tag{12.25}$$

Since

$$(\hat{l}^2 + \hat{s}^2)\phi_{njlm} = [l(l+1) + \tfrac{3}{4}]\hbar^2 \phi_{njlm} \tag{12.26}$$

it follows that

$$\hat{s} \cdot \hat{l} \phi_{njlm} = \tfrac{1}{2}[j(j+1) - l(l+1) - \tfrac{3}{4}]\hbar^2 \phi_{njlm}. \tag{12.27}$$

The effect of the operator $\hat{s} \cdot \hat{l}$ is thus seen to depend on whether $j = l \pm \frac{1}{2}$

$$\hat{s} \cdot \hat{l} \rightarrow \begin{cases} \frac{1}{2}l & j = l + \frac{1}{2} \\ -\frac{1}{2}(l+1) & j = l - \frac{1}{2}. \end{cases} \tag{12.28}$$

It will be seen that the so-called spin–orbit splitting, i.e. the energy difference between single-particle states with $j = l \pm \frac{1}{2}$, respectively, will increase as $(2l + 1)$.

It was found that by making the overall strength of the spin–orbit term in equation (12.22) large enough†, and in particular by choosing its sign so that the $j = l + \frac{1}{2}$ state lay below the $j = l - \frac{1}{2}$ state, then the observed magic numbers could be obtained, provided suitable parameters were chosen in equation (12.21) for the spin-independent term $V_{\text{SW}}(r_i)$: with R and a fixed as in §4 the best value of V_0 is about 50 MeV. The corresponding level scheme is shown in figure 12.1, where the role of the spin–orbit splitting in determining the large gaps that correspond to the magic numbers will be seen. (Standard spectroscopic notation is followed in the labelling of the states in this figure: s, p, d, f, g, h, i and j refer to $l = 0$, 1, 2, 3, 4, 5, 6

† The choice of form for the radial function $V_{\text{so}}(r_i)$ is not very critical.

and 7, respectively. Half integers denote the j quantum number, and the degeneracy of a given j state is given by $(2j + 1)$.)

This is the essence of the nuclear shell model, proposed in 1948 by Goeppert-Mayer, and independently by Haxel, Jensen and Suess. More recently the model has been generalised to include the possibility that the common field is not spherically symmetrical (it is for this reason that we have written $V(r_i, q_i)$ and not $V(r_i, q_i)$ in equation (12.6)). Also, one attempts to take account of deviations from true independent-particle motion, i.e. of correlations, essentially by treating residual interactions between the nucleons in perturbation theory. In fact, with these generalisations the shell model embraces all the other structure models that have been proposed: it is *the* nuclear model.

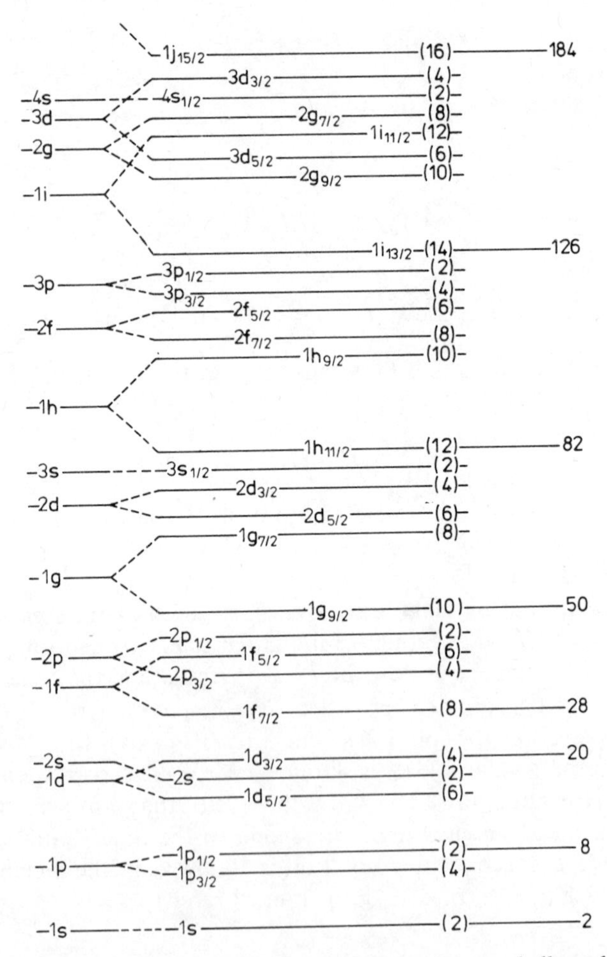

Figure 12.1 Level scheme in the Mayer–Jensen shell model.

With a suitable phenomenological determination of the parameters of the shell-model potential, the model wavefunction Ψ_0 for any given state is readily determined. A shell-model estimate can then be obtained for the expectation value of any observable O: we have for this estimate

$$\bar{O}_0 = \langle \Psi_0 | \hat{O} | \Psi_o \rangle \tag{12.29}$$

where $\hat{O}$ is the operator corresponding to the observable O. Likewise, the amplitude for a β or electromagnetic transition between two states is estimated in the shell model as

$$T_0^{\mathrm{fi}} = \langle \Psi_0^{\mathrm{f}} | \hat{\mathscr{T}} | \Psi_0^{\mathrm{i}} \rangle \tag{12.30}$$

where $\hat{\mathscr{T}}$ is the appropriate transition operator, while Ψ_0^{i} and Ψ_0^{f} are the shell-model approximations to the initial and final states, respectively (the probability of the transition is given as $|T_0^{\mathrm{fi}}|^2$). In this way the shell model has enjoyed an enormous success in correlating numerous data relating to the systematics of nuclear spins, magnetic and electric static moments, electromagnetic transition probabilities and β transition probabilities. However, there is one quantity which the shell model fails completely to reproduce, and that is the internal energy. This is particularly unfortunate from the point of view of this book, since it is precisely this property above all others which interests us.

To see what happens, we note that there are two different ways in which we can estimate the internal energy of a nucleus. Firstly, we can take the model value E_0, given by (12.14) simply as the sum of single-particle energies; for those nuclei in which all the single-particle energies are known† it is found that E is far too negative, i.e. equation (12.14) predicts a strong over-binding. For example, in the case of ^{16}O, the estimate (12.14) gives $E_0 \simeq -400$ MeV, compared to the experimental value of -127 MeV. (On the other hand, equation (12.14) gives a good account of the *variation* of the binding energy from one nucleus to another: it was in this way that the shell model was able to account for the observed magic numbers.)

Alternatively, we can follow the prescription (12.29), expressing the energy as the expectation value of the exact Hamiltonian (12.3)

$$\begin{aligned} \bar{E}_0 &= \langle \Psi_0 | H | \Psi_0 \rangle \\ &= \langle \Psi_0 | \hat{T} | \Psi_0 \rangle + \langle \Psi_0 | \hat{V} | \Psi_0 \rangle \end{aligned} \tag{12.31}$$

where $\hat{T}$ and $\hat{V}$ are given by equations (12.4) and (12.5), respectively. (Actually, our first estimate, given by equation (12.14), corresponds to

† Although the shell-model potential (12.22) can be adjusted to give the correct energies of the highest occupied single-particle states, and thus the correct magic numbers, it can never give *all* the single-particle energies. To do this it is necessary to make the potential well momentum-dependent.

using equation (12.29) but with the exact Hamiltonian H replaced by the model Hamiltonian H_0 of equation (12.7).) The estimate (12.31) always turns out to be far too *positive* i.e. there is not enough binding.

However, the overall experimental success of the shell model is evident, despite this one conspicuous failure for the case of the internal energy. Historically, though, it was not at all clear why the shell model should give such a good approximation to the real wavefunction. Indeed, the idea of independent-particle motion was at first strongly resisted, for there seemed to be absolutely no reason why the interaction experienced by a given nucleon could be represented, even as an approximation, by a potential field. Thus, at the same time as the shell model was being developed and refined phenomenologically with spectacular success, a considerable effort was being devoted to the more fundamental problem of relating it to the real nucleon–nucleon forces, i.e. to the basic Schrödinger equation (12.2), and in particular to understanding why the model worked at all. This problem really lies far beyond the scope of this book, and indeed it has not yet been completely resolved. Nevertheless, we shall now attempt to sketch in outline our qualitative understanding of the matter.

Why does the shell model work? To understand both the overall success of the shell model, and its failure in the case of the internal energy, it should first be realised that the real wavefunction Ψ of the nucleus can never take *exactly* the independent-particle form Ψ_0, i.e. there must always be *some* correlations in the wavefunction. We can see this mathematically from the fact that it is impossible to express the exact Hamiltonian H, given by equations (12.3)–(12.5), of which Ψ is an eigenstate, in the form of the Hamiltonian H_0 of equation (12.7). Physically, too, there is one obvious defect to the independent-particle state Ψ_0 of equation (12.10), and that is the fact that there is nothing in this function that prevents two nucleons from approaching each other arbitrarily closely, even though in reality they cannot, because of the strong short-range repulsion. This last point shows at once why the shell-model estimate (12.31) of the energy E is much too high: the problem arises in the expectation value of V, essentially because the model wavefunction drastically exaggerates the probability of two nucleons finding themselves within their region of strong mutual repulsion.

Having understood, then, how correlations do in fact play a vital role in the case of the energy, we see that the real question is why are the correlations so relatively unimportant from the point of view of all the other quantities we have considered? This question can be reformulated as follows. Let us write the exact wavefunction in the form

$$\Psi(r_1, \ldots, r_A) = \Psi_0(r_1, \ldots, r_A)F(r_1, \ldots, r_A) \qquad (12.32)$$

where $F(r_1, \ldots, r_A)$ is the *correlation function*: this would be just equal to unity if the shell model were exact. Then the good estimates given by the shell model for the expectation value of all quantities other than energy imply that the correlation function can be written in the form

$$F = \prod_{(i,j)} f(r_{ij}) \tag{12.33}$$

where for all pairs of nucleons (i, j) the function $f(r_{ij})$ rapidly approaches unity as $r_{ij} \equiv |r_i - r_j|$ increases beyond the radius r_c of the short-range repulsion, which is about 1 fm. The exact function Ψ would then be well approximated by Ψ_0 over some 90% of the nuclear volume.

There really seems to be no other way of reconciling the overall success of the shell model with the existence of a strong short-range repulsion in the basic nucleon–nucleon interaction. The question has thus reduced to: why does the correlation function have this 'healing' property, as Weisskopf put it.

A significant advance was made as early as 1950, when Weisskopf stressed the crucial role played by the Pauli principle in damping down the scatterings that lead to a sharing of energy and momentum between colliding nucleons. However, this cannot be the whole story since most fermion systems do not, after all, show independent-particle characteristics; at low temperatures, for example, nitrogen becomes a crystal and not a Fermi gas.

The essential factor that determines whether or not the wavefunction heals to the independent-particle form beyond the repulsive core is the strength of the interaction in the same region, i.e. the strength of the attractive tail (see figure 12.2). (We know that this tail is attractive since the overall force must be attractive, nuclei being bound.) Now a measure of the strength of this well is given by the number of bound states that it can sustain. It is significant that the nucleon–nucleon system has only one bound state, the deuteron — if we try to excite this it simply disintegrates — while the vibrational spectra of diatomic nitrogen molecules testify to the large number of bound states that can be sustained by the atom–atom potential. Thus, the fact that independent-particle motion exists in nuclei but not in solid nitrogen, for example, is attributed to the nucleon–nucleon force being much *weaker* than the atom–atom force, in the sense that it can sustain far fewer bound states. Of course, in an absolute sense the nucleon–nucleon interaction is much *stronger* than the atom–atom interaction, its depth being of the order of MeV, rather than eV, as for the latter. The reason why the nucleon–nucleon interaction nevertheless has fewer bound states is simply because its range is so much smaller†.

A lot of progress has been made on these lines towards understanding qualitatively how the success of the shell model can be reconciled with the

† We have presented here a grossly simplified account of an argument due to Gomes, Walecka and Weisskopf, *Ann. Phys., NY* **3** 241 (1958).

short range of the nucleon–nucleon force, and in particular with the presence of a strong repulsive core. Nevertheless, it has still not been possible to derive the shell-model potential $V(r_i, q_i)$ accurately and rigorously from the nucleon–nucleon force, essentially for the same reasons that the exact Schrödinger equation (12.2) cannot be solved with any accuracy for nuclei with more than three or four nucleons.

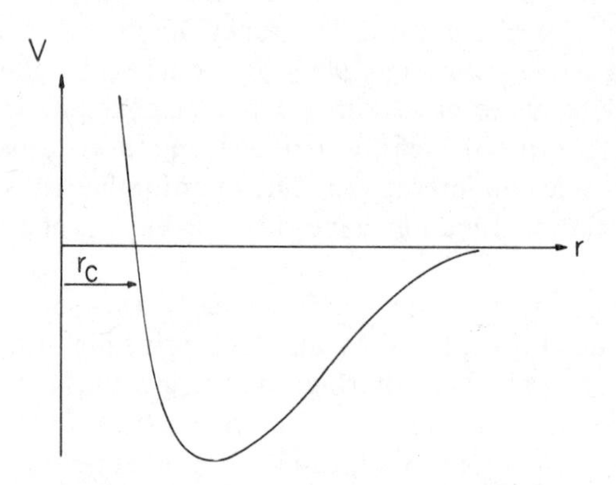

Figure 12.2 Schematic appearance of an interparticle interaction potential that saturates.

Calculation of energy. Our rapid survey of the shell model is now complete, but we see that our original problem of determining the binding energy of nuclei from the nucleon–nucleon force is still unsolved. Not only is it impossible to solve the exact Schrödinger equation (12.2), but even the shell model does not help us, since we cannot use the empirical shell-model states to calculate the energy from the expectation value (12.31) of the exact Hamiltonian, on account of the importance of the correlations.

One possible way in which the shell model might still be used to calculate energies is to compensate for the omitted correlations by replacing the real nucleon–nucleon force with some effective force. This is the essence of the nuclear Hartree–Fock method, and it is to be noted that in calculating the energy this method also calculates the form of the shell-model potential. However, this method is quite complicated and at the same time it is not completely rigorous, since a phenomenological approach has to be followed in defining the effective nucleon–nucleon force: it cannot be derived exactly from the real force.

Thus, rather than use the shell model in this way, we shall follow in this book the much simpler approach of using models, of necessity very crude, that give us directly the expectation values of $\hat{T}$ and $\hat{V}$, i.e.

$$\bar{T} = \langle \Psi | \hat{T} | \Psi \rangle \tag{12.34a}$$

and

$$\bar{V} = \langle \Psi | \hat{V} | \Psi \rangle. \tag{12.34b}$$

Insofar as we are interested only in binding energies we do not need the wavefunction at all. We present this approach in the next two sections.

§13 SATURATION: NUCLEAR MATTER

As a guide to the formulation of our crude models for $\hat{T}$ and $\hat{V}$ we recall from §12 how the simple fact that all nuclei have more or less the same density implies that nucleons can interact only with their nearest neighbours. In other words, the number of bonds a given nucleon can form is fixed, and in particular cannot increase as the total number of nucleons A increases: we say that the nuclear forces are *saturated*. One consequence of this is that the mean potential energy per nucleon is independent of A, so that the mean potential energy of the whole nucleus will be proportional to A

$$\bar{V} \propto A. \tag{13.1}$$

This, then, is our crude model for $\bar{V}$, although we shall refine it somewhat in the next section. We would stress that the saturation property on which it is based is a consequence both of the short range of the nuclear forces and of the existence of a very short-range repulsive component: it is the latter that holds the density constant.

We remarked in §4 that the constant density of nuclei is reminiscent of the behaviour of ordinary matter. The analogy is in fact closer, since the energy of a piece of ordinary matter is proportional to the number of particles it contains, as in equation (13.1) for nuclei. The reason is, of course, the same: not only do interatomic, or intermolecular, forces become repulsive at very small distances of separation, thereby ensuring constant density, but they are short ranged, too. That is, these forces, like nuclear forces, are saturated.

It is appropriate at this point to digress briefly and reflect on the origin of interatomic, or intermolecular forces. Here it is the ordinary electric force that is responsible: unlike the situation prevailing within a nucleus, both signs of charge are present and in particular there will be an attraction between the electrons of one atom and the positively charged nucleus of another. Of course, this attraction will tend to be cancelled by the repulsion between the nuclei of the two atoms and also between their respective electron clouds. In fact, this cancellation would be exact if the electron cloud around the nucleus in each atom remained spherically symmetrical, so we must suppose that when an interaction *does* arise between two atoms the electron cloud of one is being deformed by the presence of the other. The net force between the atoms will then be the resultant of the incomplete cancellation of repulsive and attractive electrostatic forces acting between equal numbers of pairs of like and unlike charges, respectively. It will be easy to understand that even though the force between a given pair of charges obeys the Coulomb law, the resultant interaction potential will, because of this cancellation, fall off much faster than $1/r$.

To return to our main argument, the analogy between ordinary matter and nuclei suggests at once the concept of *nuclear matter*, different nuclei just consisting of differently sized pieces of this substance. Of course, the analogy is not complete, the most striking difference being that, unlike ordinary matter, there appears to be a very definite upper limit on the size of a piece of nuclear matter that can exist (see, however, §15). Nevertheless, the concept of nuclear matter has proved to be a most useful fiction in the development of nuclear theory, as we shall discuss at the end of §14.

So far we have not considered the internal kinetic energy of the nucleons of our nucleus, i.e. their kinetic energy in a frame in which the centre of mass is at rest. To study this we adopt the simplest possible model of a nucleus: that of non-interacting nucleons contained in a box with rigid walls, corresponding to the nuclear surface. This picture really implies an extreme form of the independent-particle model, with an infinitely deep square well as the potential field (see, however, p. 68). Classically, the internal kinetic energy in this ideal-gas model is given simply by the thermal energy kT, which would be utterly negligible for nuclei in thermal equilibrium with their environment at ordinary temperatures. However, while the classical viewpoint is certainly valid for ordinary matter, the great density of nuclear matter means that it is essentially a quantum system, so that Maxwell–Boltzmann statistics are quite inapplicable.

The nucleon gas of our model obeys rather Fermi–Dirac statistics. We analyse the Fermi gas in some detail in Appendix C, but before we can use the results given there we have to be sure that our Fermi gas is completely degenerate, i.e. is in its ground state. For this to be the case equation (C.14) must hold, i.e. the temperature T (this is not to be confused with the kinetic energy) must satisfy the condition

$$kT \ll \varepsilon_F \tag{13.3}$$

where ε_F is the Fermi energy. Now in a nucleus neutrons and protons form distinct Fermi systems, so each should have its own Fermi energy. To simplify matters let us suppose that $N = Z$, so that the two Fermi energies are equal; this is never far from the case anyway. Then equation (C.11) becomes

$$\varepsilon_{\mathrm{F}}=\frac{\hbar^2}{2M}\left(\frac{3\pi^2 A}{2V}\right)^{2/3}=\frac{\hbar^2}{2M}\left(\frac{9\pi}{8}\right)^{2/3}\frac{1}{r_0^2} \tag{13.4}$$

where we have used equation (4.2) for the volume V of the nucleus. The value $r_0 \simeq 1.15$ fm then gives $\varepsilon_{\mathrm{F}} \simeq 36$ MeV. The corresponding temperature is 4×10^{11} K, so that under all normal terrestrial circumstances nuclei in thermal equilibrium with the environment will be in the ground state. This does not mean that a nucleus in these circumstances cannot be excited, but rather that if it is then it cannot be regarded as being in thermal equilibrium, and will tend to return to the ground state by emission of a photon (γ decay). However,under the extreme temperatures of the hottest stars the condition (13.3) will not be satisfied, and even at thermal equilibrium a certain fraction of nuclei will be excited (the excitation is maintained by the very high-energy photons of the black-body radiation, and by collisions between the nuclei).

With the results of Appendix C thus being applicable to the nucleus, we see from equations (C.11) and (C.13) that the mean kinetic energy per nucleon will be determined by the density and will thus be the same for all nuclei. Thus the internal kinetic energy of a nucleus, like the potential energy, will be proportional to the number of nucleons, so that we shall have a relation comparable to equation (13.1)

$$\bar{T} \propto A. \tag{13.5}$$

It follows that we can write for the total internal energy

$$E \propto A$$

i.e.

$$E = -a_{\mathrm{vol}}A. \tag{13.6}$$

The value of $-a_{\mathrm{vol}}$ is found experimentally (see §14) to be about -16 MeV: this represents the internal energy per nucleon of nuclear matter. We call this the *nuclear-matter approximation* to the internal energy of nuclei, since such a proportionality is an essential characteristic of ordinary matter†. However, significant deviations from proportionality occur for nuclei, because of severe limitations in the nuclear-matter concept that underlies our crude models for $\bar{T}$ and $\bar{V}$, as will be discussed in the next section.

Referring once again to equations (C.11) and (C.13), we see that the kinetic energy per nucleon increases as the 2/3 power of the density. The associated compressional pressure thus resists to some extent the tendency

† It will become clear in §14 why we regard this approximation to E as a volume term.

to collapse that would exist if the nucleon–nucleon forces were purely attractive. However, it may be shown that the potential energy per nucleon would in this case vary linearly with the density, at least at high densities, so that the kinetic energy alone can never stop the collapse and it is indeed necessary to invoke a repulsive component in the force, as was tacitly assumed in the discussion at the end of §4. The actual density of nuclear matter will be determined by the equilibrium condition

$$\frac{\mathrm{d}E}{\mathrm{d}\rho} \equiv \frac{\mathrm{d}\overline{T}}{\mathrm{d}\rho} + \frac{\mathrm{d}\overline{V}}{\mathrm{d}\rho} = 0 \tag{13.7}$$

this just minimising the total energy with respect to the density ρ.

Now with ε_F taking the value of about 36 MeV it follows from equation (C.13) that the mean kinetic energy per nucleon, $\bar{t}_{vol}$, is about 21 MeV, which is incomparably larger than in the classical picture, and far from negligible. Since the total internal energy per nucleon in nuclear matter is, as we have stated, about -16 MeV, it follows that the mean potential energy $\bar{v}_{vol}$ must be about -37 MeV per nucleon: more than half of this is cancelled out by the kinetic energy.

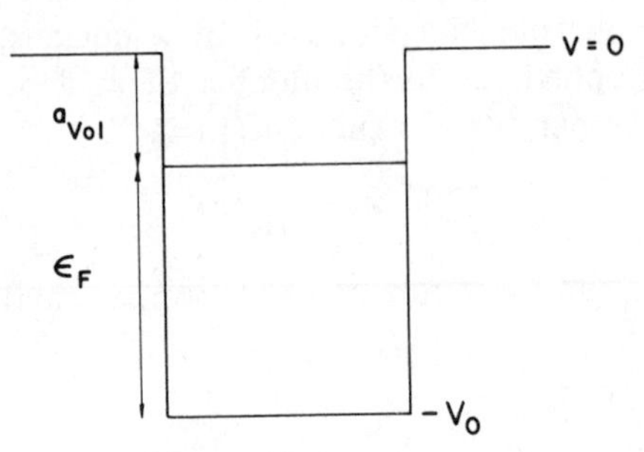

Figure 13.1 Potential well for Fermi-gas model of the nucleus: separation energy corresponds to infinite nuclear matter.

The following inconsistency in our Fermi-gas model now becomes apparent. The necessary assumption that the walls of the container are rigid means that the equivalent shell-model potential is infinitely deep. But then an infinite amount of energy would be required to remove a nucleon from the nucleus, even one with the maximum kinetic energy of $\varepsilon_F \simeq 36$ MeV. However, with $-a_{vol}$ denoting the mean energy per nucleon, it follows that the necessary energy to remove a nucleon is, in fact, just $a_{vol} \simeq 16$ MeV†.

† This is the nucleon separation energy S_N (§17) for nuclear matter. However, in finite nuclei, for which surface, Coulomb and symmetry effects (§14) have to be taken into account, S_N varies between 11 MeV in light nuclei and 5 MeV in heavy nuclei (Problem III.13).

Thus the Fermi surface must be located at a depth of $-a_{vol}$, from which, referring to figure 13.1, we see that the depth of the equivalent well must be

$$V_0 = a_{vol} + \varepsilon_F$$
$$\simeq 52 \text{ MeV}.$$

We have already remarked (§12) that this is the well depth of the Saxon–Woods potential of equations (12.21) and (12.22) required to get the correct magic numbers.

The fact that the equivalent well is not infinitely deep will have some repercussions on our calculation of the kinetic energy. The effect will manifest itself as a kinetic energy contribution to the surface correction (see §14).

§14 SEMI-EMPIRICAL MASS FORMULA

In this section we refine somewhat the crude models for kinetic and potential energies that underlie our so-called nuclear-matter approximation (13.6) to the internal energy of the nucleus. For simplicity the quantity that we shall consider is the internal energy per nucleon

$$e = \frac{E}{A} = -\frac{B}{A} \tag{14.1}$$

which is, of course, negative, and the more negative it is the more stable the nucleus. According to equation (13.6), this quantity is the same for all nuclei

$$e = -a_{vol} \tag{14.2}$$

but in fact significant deviations from strict constancy are found, and our main preoccupation in this section will be with representing these deviations.

Let us begin by summarising the main trends in these deviations. In general, e is a function both of N and Z, i.e. $e = e(Z, N)$ or $e(A, Z)$, and we shall have to represent it by a surface in a three-dimensional plot, but at first we shall limit ourselves to considering only the most strongly bound isobar for each value of the mass number A. That is, for each value of A we consider only the atomic number Z_0 that satisfies

$$\left(\frac{\partial e(A, Z)}{\partial Z}\right)_A = 0. \tag{14.3}$$

We have already remarked on the phenomenon of β radioactivity, i.e. the spontaneous transmutation of a nucleus to a more strongly bound nucleus having the same value of A. Nuclei satisfying equation (14.3) will clearly be stable with respect to this isobaric process (in fact, because of the slight difference between the neutron and proton masses this may not always be exactly true; see §16), so that the function $Z_0(A)$ they define is referred to as the *line of β stability* (or simply as the 'stability line', although it should be realised that the nuclei lying thereon are not necessarily stable with respect to the other forms of disintegration that we discuss in Chapter III, namely particle radioactivity and spontaneous fission). Figure 14.1 shows this curve plotted in the N–Z plane; we see in particular the tendency towards equality of Z and N in light nuclei, together with the growing neutron excess as we move towards heavier nuclei. Limiting ourselves, then, to nuclei that lie on this stability line, e will be a function of A only, and so can be represented in a two-dimensional plot; the experimental values fall close to the full curve shown in figure 14.2.

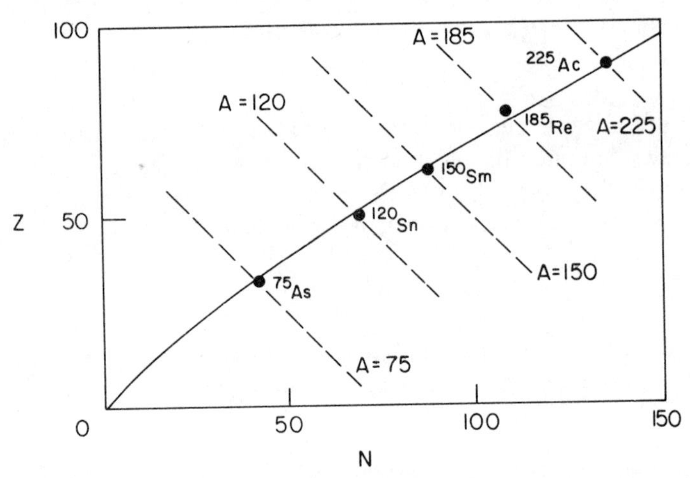

Figure 14.1 The line of β stability (full curve). Nuclei lying on the broken cross lines are all isobars belonging to the indicated values of A; the nuclei shown are the most stable for these values of A.

We see in fact that the variation in e along the stability line is quite weak, most nuclei having a binding energy of around 8 MeV per nucleon†. Nevertheless there is a definite trend, and we see that there is a minimum in this curve at $A = 56$: the corresponding nuclide, ^{56}Fe, thus has the greatest binding energy per nucleon and must be regarded as the most stable nuclide

† Although this approximate constancy in e immediately suggests the nuclear-matter approximation (14.2), we shall see that the numerical value of a_{vol} is quite different from 8 MeV—see equation (14.19).

(the implications of this for stellar nucleosynthesis will be discussed in Chapter VI and Appendix D). The binding energy per nucleon falls off for both A larger than 56 and also for smaller values.

We can discuss the binding-energy systematics for nuclei *off* the stability line in terms of the variation along lines of constant A, such as are shown in figure 14.1. However, beyond making the obvious statement that there is a minimum at $Z = Z_0$ for each such line, i.e. on the stability line, we shall postpone our discussion of this isobaric variation of e until we are ready to analyse it.

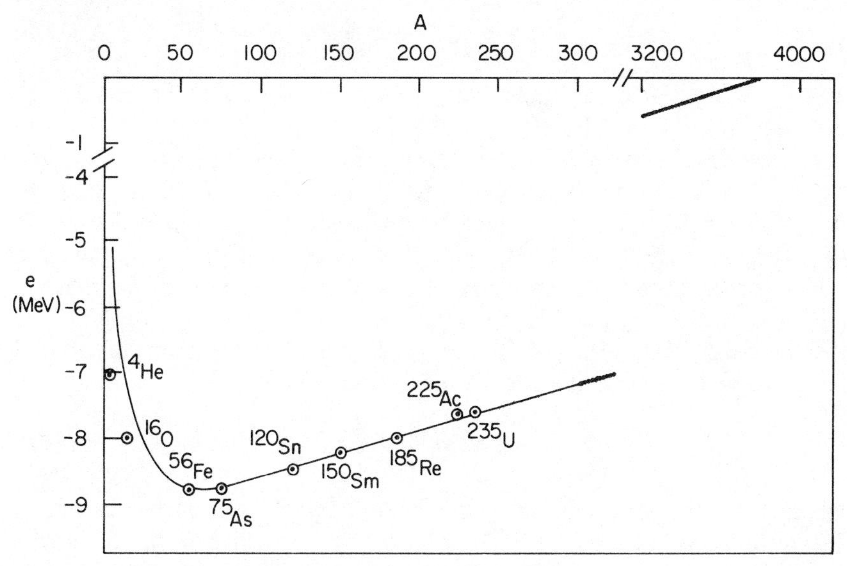

Figure 14.2 The internal energy (= − binding energy) per nucleon along the stability line, according to equation (14.12).

We shall see in the following that with a very small number of extremely simple corrections to equation (14.2) we obtain a formula, the so-called *semi-empirical mass formula*, which represents fairly accurately all the main features of the variation of e with Z and N, provided the three or four constants appearing in the formula are appropriately adjusted.

Two obvious limitations of our nuclear-matter model are: first that nuclei are finite rather than infinite, so that surface effects should arise; second nuclei are charged, and the Coulomb interaction between protons, being long-ranged, is non-saturating. We shall discuss the appropriate corrections for these two effects immediately, and then see what further refinements are necessary.

Surface term. The phenomenon of surface tension shows us that even in ordinary matter the binding energy of a sample is not strictly proportional to the number of particles but is *reduced* by an amount proportional to the surface area of the sample. More precisely, for an incompressible liquid drop we have for the total internal energy

$$E = -b_{vol}V + b_{surf}S \tag{14.4}$$

where V is the volume of the drop, S its surface area and both parameters are positive. Regarding, then, the nucleus as a liquid drop, and taking note of the relation (4.2) for the radius, we shall have for a nucleus of mass number A

$$e = -a_{vol} + a_{surf}\frac{1}{A^{1/3}} \tag{14.5}$$

where a_{vol} and a_{surf} are positive parameters that will, of course, be eventually determined by fitting to the data.

In the case of a liquid drop, the reduction in the binding energy associated with the surface comes from the fact that a particle near the surface has fewer close neighbours with which to interact than does a particle deep in the interior of the liquid. A simple calculation, the details of which can easily be reconstructed from figure 14.3 (see Problem II.6), shows that the potential energy per particle $\bar{v}$ is reduced by a factor $[1-9r_n/(16R)]$, so that in the nuclear case, with the radius R given by (4.2), we shall indeed be led to a relation of the form (14.5), with

$$a_{surf} = -\frac{9}{16}\frac{r_n}{r_0}\bar{v}_{vol} \tag{14.6}$$

where $\bar{v}_{vol}$ is the value of $\bar{v}$ in infinite nuclear matter, i.e. in the absence of surface effects. Since $\bar{v}_{vol} \simeq -37$ MeV† we have $a_{surf} \simeq 21(r_n/r_0)$ MeV, the exact value of which depends on what we take for r_n, but which in any case is surprisingly close to the experimental value of about 20 MeV (see equation (14.19)). However, this agreement may be somewhat fortuitous, since there should also be a surface correction associated with the kinetic energy in the case of dense systems such as nuclei (see §13).

It will be obvious now why the 'nuclear-matter' term $-a_{vol}$ is regarded as a volume or bulk term.

Coulomb term. The argument that nucleons interact only with their nearest neighbours depends on the short range of the force and thus cannot apply to the Coulomb interaction between protons. Rather, each proton interacts with *all* other protons, at least as far as the Coulomb force is concerned, so that the saturation property is not satisfied, the electrostatic

† Remember that our derivation of this value in §13 depended upon our knowledge of the experimental value of a_{vol} (see equation (14.19)).

potential energy E_C being proportional not to the number Z of protons but instead to the number $\frac{1}{2}Z(Z-1)$ of proton pairs†. For large Z this means that E_C varies as Z^2, and in fact if we assume that the nuclear charge Ze is smeared out uniformly over the nucleus, we have the well known result of classical electrostatics (see Problem II.7)

$$E_C = \frac{3}{5}\frac{Z^2e^2}{R}. \tag{14.7}$$

Then substituting from (4.2) for the nuclear radius, the Coulomb correction to the energy per nucleon takes the form

$$e_C = a_C \frac{Z^2}{A^{4/3}}. \tag{14.8}$$

It follows that our earlier value for r_0, 1.15 fm, would give $a_C \equiv 3e^2/5r_0 = 0.751$ MeV.

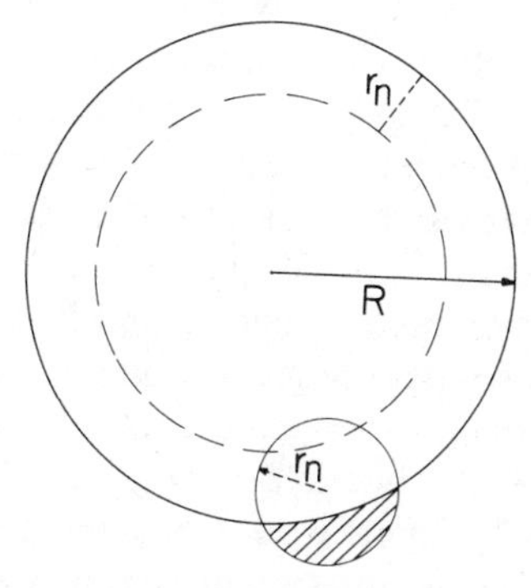

Figure 14.3 Calculation of surface correction to potential energy. All particles lying in the surface shell of thickness r_n (range of force) will have their 'volume of effective interaction' reduced by an amount corresponding to the hatched cap of the sphere of radius r_n.

However, the use of (14.7) ignores the fact that the nuclear surface is not perfectly sharp, and that this value of r_0 represents the density only at the centre of the nucleus (note that the non-saturating effect of the Coulomb force is not strong enough to destroy the approximate constancy of this density as we go from one nucleus to another). Furthermore, (14.7) is a purely classical expression, and quantum mechanics will make some difference, mainly through the action of the Pauli principle. Since we are not

† Note that this problem of non-saturation does not arise for the interaction between atoms in ordinary matter, even though this interaction is Coulombic in origin; see the discussion in §13.

taking any of these factors into account explicitly we should determine a_C empirically from nuclear-mass systematics, and may do this most simply by comparing mirror nuclei, i.e. pairs of nuclei whose relationship to each other consists of all neutrons being replaced by protons and vice versa (§8). Then because of charge symmetry of nuclear forces the energy difference between two such nuclei must lie entirely in the Coulomb term (14.7) (see the form (14.12) of the mass formula). In this way (see Problem II.8) we find

$$a_C = 0.60 \text{ MeV} \tag{14.9}$$

which corresponds to an effective value of 1.44 fm for r_0.

Thus we now have in place of (14.5) for the internal energy per nucleon

$$e = -a_{\text{vol}} + a_{\text{surf}} \frac{1}{A^{1/3}} + a_C \frac{Z^2}{A^{4/3}} \tag{14.10}$$

where a_C is given by (14.9).

Recalling the general increase of Z with A, we see that equation (14.10) reproduces, at least qualitatively, the variation of e along the stability line shown in figure 14.2. The decrease in e, i.e. increase in binding energy per nucleon, that occurs with increasing A in light nuclei is a surface effect, while the subsequent increase in e for heavy nuclei is a Coulomb effect. Nevertheless, (14.10) does not tell us why Z should increase with A. Rather, for a given value of A the binding energy, and hence the stability, would be greater for smaller Z, so that all nuclei would tend to β decay to the appropriate isotope of hydrogen—remember that the factor Z^2 in (14.10) should really read $Z(Z-1)$. Thus hydrogen would be the only stable element. On the other hand, there would be no upper limit on the mass number, the binding energy per nucleon increasing steadily with A (towards the limiting value of a_{vol}), on account of the decreasing importance of the surface energy.

Symmetry term. The obviously unacceptable state of affairs arising in the above discussion can be rectified by postulating the existence of an additional correction term that tends to equalise N and Z. The simplest form of such a term compatible with charge symmetry of the nuclear forces is a quadratic dependence on the variable

$$I = \frac{N-Z}{A} \tag{14.11}$$

which represents the fractional neutron excess, so we replace (14.10) by the expression

$$e = -a_{\text{vol}} + a_{\text{surf}} \frac{1}{A^{1/3}} + a_C \frac{Z^2}{A^{4/3}} + a_{\text{sym}} \left(\frac{N-Z}{A} \right)^2 \tag{14.12}$$

which is known as the mass formula of Bethe and von Weizsäcker. We postpone a discussion of the origin of the new term, the so-called *symmetry term*†, until after we have examined its phenomenological implications.

First, we see that (14.12) gives the correct form for the stability line, and in particular there will no longer be the tendency for all nuclei to β decay down to hydrogen. Using (14.3), and noting that $N = A - Z$, we have for the β stable isobar

$$Z_0 = \frac{A}{2 + xA^{2/3}} \tag{14.13}$$

where

$$x = \frac{a_C}{2a_{sym}}. \tag{14.14}$$

We see in (14.13) the tendency towards equality of N and Z for light nuclei, together with the growing neutron excess for increasing mass number. Effectively, the Coulomb term and symmetry term are competing in the attempt to minimise the energy, with the latter tending to keep Z equal to $A/2$, and the former trying to make Z as small as possible. Now although both terms are quadratic in Z the Coulomb term varies as $A^{-4/3}$ while the symmetry term varies as A^{-2}, so that the former will become steadily more influential with increasing A, in the sense that it will force Z more and more below the value $A/2$. Somewhat paradoxically, this means that although for small A the symmetry term is much smaller than the Coulomb term it grows faster than the latter with increasing A, simply because of the slow increase in Z and the consequently more rapid growth of the asymmetry I. We discuss this point in more detail below.

The actual value of a_{sym} can be found by fitting x in (14.13) to Z_0 for several different values of A. We find $x = 0.014$, the resulting curve being the one shown in figure 14.1; the fit, obtained with just one parameter, is remarkably good. Then with the value (14.9) for a_C we have

$$a_{sym} = 21.4 \text{ MeV}. \tag{14.15}$$

Substituting (14.13) into (14.12) for the energy per nucleon for nuclei lying on the stability line gives

$$e = -a_{vol} + a_{surf}\frac{1}{A^{1/3}} + a_C\frac{1}{2x + 4A^{-2/3}}. \tag{14.16}$$

Such an expression always gives a curve of the form shown in figure 14.2, with just one minimum, whatever the values of the coefficients. The

† For a given neutron–proton composition, I, the contribution of this term to the total internal energy of the nucleus varies as A. Thus it is just as much a 'nuclear-matter' or 'volume' term as the one in a_{vol}.

decrease in e that occurs at first with increasing A is associated with the surface term, as we remarked when commenting on equation (14.10). However, the subsequent increase in e is more a joint Coulomb-symmetry effect rather than a pure Coulomb effect; we shall return to this point below. The position of the minimum ($A = 56$) is given by

$$a_{\text{surf}} = \frac{2a_C A}{(2 + xA^{2/3})^2} = 13.9 \text{ MeV}. \tag{14.17}$$

The one remaining parameter, a_{vol}, whose sole effect is simply to shift the curve vertically without affecting its form, is chosen to give the best overall fit; the value that we shall take, and to which the curve in figure 14.2 corresponds, is

$$a_{\text{vol}} = 14.4 \text{ MeV}. \tag{14.18}$$

The quality of the fit of equation (14.12) to the masses of β stable nuclei is excellent, considering that only two parameters were available, x and a_C being already determined. We shall see below that (14.12) also reproduces the main trends for nuclei lying off the stability line, so that we shall regard it as our final form of the semi-empirical formula, except insofar as we shall later add a fine-structure term. However, although the Bethe–von Weizsäcker formula (equation (14.12)) does exceptionally well, considering its simplicity, in reproducing the main systematics of nuclear binding energies, more complicated formulae have been proposed in an attempt to improve still further the fit to individual nuclei. These formulae require a considerable change in the numerical values of the above coefficients

$$a_{\text{vol}} \simeq 16 \text{ MeV} \qquad a_{\text{surf}} \simeq 20 \text{ MeV} \qquad a_{\text{sym}} \simeq 30 \text{ MeV} \qquad r_0 \simeq 1.15 \text{ fm} \tag{14.19}$$

(note that the latter is consistent with the value determined by electron scattering). Although these must be regarded as more reliable estimates of the physical quantities in question, we emphasise that in using the formula (14.12), the values (14.18), (14.17) and (14.15), respectively, should be taken, along with the value (14.9) for a_C.

With the increase in e being monotonic beyond $A = 56$, it will eventually become positive; we see in figure 14.2 that this happens at around $A = 3800$ (see Problem II.10). Nuclei heavier than this will have no binding energy at all and hence cannot possibly exist (see, however, §15). However, long before this limit is reached the decreasing binding energy has the effect of making nuclei so unstable with respect to spontaneous fission (see Chapter III) that they are never observed in practice; that is why the curve in figure 14.2 is shown as dotted beyond $A = 300$.

It is most instructive to analyse the contributions of the various terms of (14.12) to the variation of e along the stability line. Referring to table14.1, we notice once again that the increase in e, i.e. loss in binding, that occurs as we move from ^{56}Fe towards lighter nuclei is indeed a surface effect. Turning to the joint Coulomb-symmetry effect that is responsible for the increase in e in the opposite direction, we note that the symmetry term is indeed small compared with the Coulomb term. Nevertheless, it is misleading to say, as we did in commenting on equation (14.10) before the introduction of the symmetry term, that this increase is essentially a Coulomb effect, to be contrasted with the surface effect that dominates in light nuclei. In fact, we see from table 14.1 that the increase in the Coulomb term between ^{56}Fe and ^{235}U is almost exactly cancelled out by the continuing decrease of the surface term†. Thus the increase in e towards the heavy nuclei depends essentially on the increase in the symmetry energy. Furthermore, the Coulomb energy is itself a direct consequence of the symmetry term, forcing, as it does, Z to follow an increase in A. Thus it is seen that the symmetry term plays a role that is completely out of proportion to the smallness of its absolute value. Of course, it must be realised that the smallness of this term comes not from the smallness of the coefficient a_{sym}, but rather from the near equality of N and Z. In fact, as already pointed out above, the symmetry term grows faster than the Coulomb term with increasing A, and on the β stability line the two terms become equal at $A = (2/x)^{3/2} \simeq 1700$, for which value of A nuclei are still bound, although highly unstable.

Table 14.1 Contributions of the variable terms in the semi-empirical mass formula (14.12) for nuclei lying on the stability line (in MeV per nucleon).

	Surface $a_{\text{surf}}A^{-1/3}$	Coulomb $a_C Z^2 A^{-4/3}$	Symmetry $a_{\text{sym}}[(N-Z)/A]^2$
^{16}O	5.51	0.95	0.00
^{56}Fe	3.63	1.90	0.11
^{235}U	2.26	3.51	1.01

As a final reminder of the significance of the symmetry term, we recall that without it all nuclei would β decay down to some isotope of hydrogen. (However, there would be no limit on the mass number, the binding energy per nucleon increasing steadily with A.) Thus the symmetry term is indispensable to the existence of stable isotopes for all elements heavier than

† It is this approximate constancy in the sum of the surface and Coulomb terms, together with its large value, that accounts for the difference between a_{vol} and the approximately constant binding energy of about 8 MeV per nucleon for all but the lightest nuclei.

hydrogen, and thus of matter as we know it. But because of the Coulomb term it is inevitable that there will eventually be a loss of binding with increasing A, and hence an upper limit on the atomic number Z. There seems to be no way in which the laws of nuclear physics could be modified to allow an indefinite range of chemical elements (if the Coulomb interaction were to be switched off there would be no atoms at all!).

Origin of the symmetry term. We shall now show how both the kinetic energy and the potential energy of the nucleus depend on the fractional neutron excess I defined in equation (14.11). Considering the kinetic term first, we note that our derivation of the mean kinetic energy per nucleon, based on the Fermi-gas model of Appendix C, assumed equal densities of neutrons and protons, and clearly has to be modified. Defining first a separate Fermi energy for neutrons and protons, we have in an obvious notation, using (C.11)

$$\varepsilon_{FN} = \frac{\hbar^2}{2M}(3\pi^2 n_N)^{2/3} \tag{14.20a}$$

and

$$\varepsilon_{FZ} = \frac{\hbar^2}{2M}(3\pi^2 n_Z)^{2/3} \tag{14.20b}$$

where

$$n_N = \frac{N}{A} n = \frac{n}{2}(1+I) \tag{14.21a}$$

and

$$n_Z = \frac{Z}{A} n = \frac{n}{2}(1-I). \tag{14.21b}$$

The mean kinetic energy per nucleon is then, according to (C.13)

$$\begin{aligned}\bar{t} &= \frac{3}{5}\left(\frac{N}{A}\varepsilon_{FN} + \frac{Z}{A}\varepsilon_{FZ}\right)\\ &= \frac{3}{5}\frac{\hbar^2}{2M}\left(\frac{3\pi^2 n}{2}\right)^{2/3}\frac{1}{2}\,[(1+I)^{5/3} + (1-I)^{5/3}]\\ &= \bar{t}_0\left(1 + \frac{5}{9}I^2 + O(I^4)\right)\end{aligned} \tag{14.22}$$

where $\bar{t}_0$ is simply the mean kinetic energy per nucleon in the original symmetric situation, $N = Z$, having a value of about 21 MeV. Then with the symmetry coefficient defined as in (14.12), we find that its kinetic energy part is

$$a_{\text{sym-kin}} = \frac{5}{9}\bar{t}_0 \simeq 12 \text{ MeV}. \tag{14.23}$$

This accounts for somewhat less than half of the experimental value (14.19), so we must obviously look for some potential energy contribution. Now the kinetic energy component was basically a consequence of the Pauli principle, and more particularly of the fact that it applies only to neutrons and protons separately, the two certainly being distinguishable. We shall see now that a potential energy contribution to the symmetry term arises from essentially the same source.

Suppose for simplicity that the interaction between two nucleons is so short-ranged that it is effective only when the nucleons are in contact. Then like nucleons will interact only when their spins are antiparallel, while a neutron and a proton will be able to interact whether their spins are parallel or antiparallel. Thus the average interaction between a neutron and a proton in nuclear matter will be twice as strong as that between two like nucleons†.

Consider now a given proton. The numbers of protons and neutrons with which it comes into contact are proportional to Z/A and N/A, respectively (we are assuming $Z, N \gg 1$), so that its mean interaction energy can be written as

$$\nu_p = V_0 \frac{Z+2N}{A} \tag{14.24a}$$

where V_0 is a constant of proportionality that evidently represents the mean interaction energy of a nucleon in the case where all nucleons are alike. Likewise the mean interaction energy of a neutron can be written as

$$\nu_n = V_0 \frac{N+2Z}{A} \tag{14.24b}$$

so that the mean interaction energy per nucleon in a system of Z protons and N neutrons will be

$$\begin{aligned} \bar{\nu}(Z,N) &= \frac{1}{2A}(Z\nu_p + N\nu_n) \\ &= \frac{V_0}{2A^2}(Z^2+N^2+4NZ) = \frac{3V_0}{4}(1-\tfrac{1}{3}I^2) \end{aligned} \tag{14.25}$$

where the factor $\frac{1}{2}$ is introduced to prevent us from counting the interaction energy of each pair of nucleons twice. Putting $Z=N=A/2$ we have

$$V_0 = \frac{4}{3}\bar{\nu}\left(\frac{A}{2}, \frac{A}{2}\right) \tag{14.26}$$

† We are ignoring here a possible spin dependence of the neutron–proton force. Actually, we know from scattering data and the binding energy of the deuteron that there is such a dependence (see Problem I.12), and when it is taken into account in more sophisticated calculations the agreement with the experimental value of a_{sym} is improved still further.

where $\bar{v}\ (A/2, A/2)$ is just the mean potential energy per nucleon in nuclear matter for the usual symmetric situation of equal densities of neutrons and protons; this is the quantity $\bar{v}_{\mathrm{vol}}$ that we calculated as approximately − 37 MeV in §13. Thus

$$\bar{v}\ (Z, N) = \bar{v}_{\mathrm{vol}}(1 - \tfrac{1}{3}I^2) \tag{14.27}$$

so that the potential energy part of the symmetry coefficient is†

$$a_{\mathrm{sym-pot}} = -\tfrac{1}{3}\bar{v}_{\mathrm{vol}} \simeq 12\ \mathrm{MeV}. \tag{14.28}$$

Combining this with (14.23) we find a considerable improvement with respect to the experimental value (14.19).

The symmetry energy in both its kinetic and potential energy parts is thus seen to be essentially a consequence of the Pauli principle. Then bearing in mind our earlier remarks about the importance of the symmetry energy for the existence of charged nuclei, we realise that if the Pauli principle were not valid there would be no chemical elements heavier than hydrogen, so that matter as we know it would be impossible‡.

Isobaric variation. Our semi-empirical mass formula (14.12) is valid for nuclei of arbitrary N and Z, but so far we have limited our attention to nuclei lying on the β stability line, and in fact the coefficients of the formula have been completely determined by these nuclei. It is now of obvious interest to turn to nuclei lying off the stability line, noting that we have no more adjustable coefficients at our disposal.

The simplest way of discussing the systematics of binding energy off the stability line is to examine its variation along lines of constant A, such as are shown in figure 14.1. Putting $N = A - Z$ in (14.12), we see that e must vary quadratically with Z for constant A. Then since there is a minimum at $Z = Z_0$, i.e. on the stability line, where we write $e = e_0$, this variation takes the parabolic form§

$$e = e_0 + f(Z - Z_0)^2 \tag{14.29}$$

in which we can determine f simply by calculating $(\partial^2 e/\partial Z^2)_A$ from (14.12);

† This result is exact for a spin-independent δ function force calculated in first-order perturbation theory.

‡ Even if we took the existence of conventional nuclei for granted, atomic structure and hence chemistry would be quite different without the Pauli principle. Thus the existence of ordinary matter depends on the operation of the Pauli principle at two different levels.

§ This explains the very rapid reduction in β lifetimes as we move away from the stability line, the parabolic variation in internal energy seen in figures 14.4 and 14.6 implying that the β decay energy increases linearly.

we find

$$f=\frac{2a_{\text{sym}}}{AZ_0}. \tag{14.30}$$

We shall see in the discussion on pairing energy below that a special effect arises in the case of even-A nuclei, so that our comparison with experiment must for the present be limited to odd-A nuclei. Referring to figure 14.4, where we show two different odd-A values, it is clear that the experimental

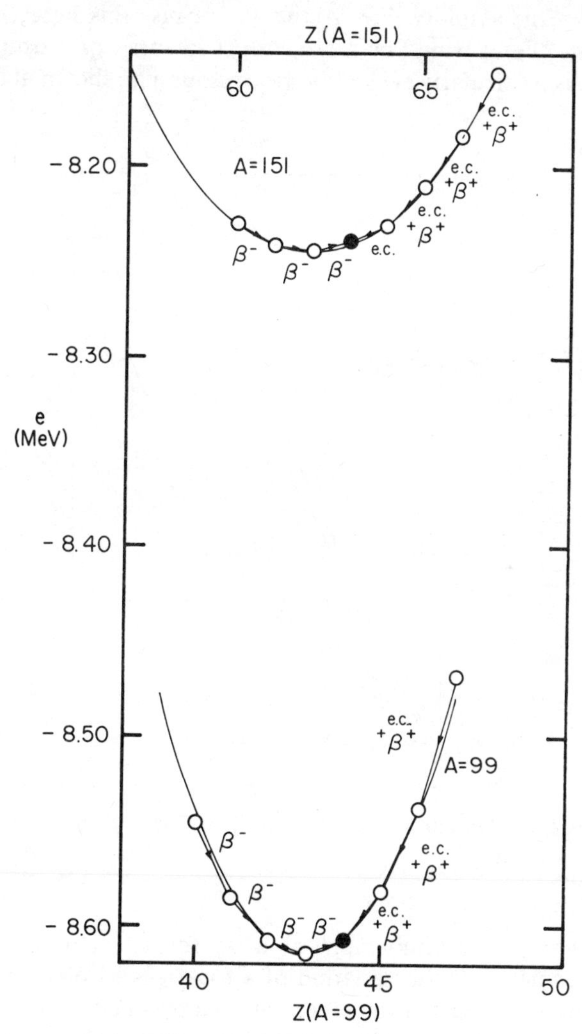

Figure 14.4 Isobaric variation of binding energy of odd-A nuclei. The arrows (β transitions) are explained in §16; the full circles represent the β stable nuclei.

energies do indeed follow a parabolic variation, with the 'steepness' f decreasing with increasing A. Furthermore, the value of a_{sym} that we extract from the 'best-fit' value of f, using (14.30), lies at around 20 MeV, the value that we obtain from the fit to the stability line (Problem II.12). Thus our formula (14.12) is seen to fit both nuclei lying on and off the stability line with a consistent set of parameters.

With the variation of e at constant A taking the parabolic form shown in figure 14.4, it is clear now that plotting e as a function of both Z and N will define a narrow valley, the so-called *valley of stability*, the floor of which follows the stability line. Actually, because this latter itself has a minimum (at ^{56}Fe) it would be more correct to speak of a trough than of a valley: this is particularly evident in the contour map shown in figure 14.5.

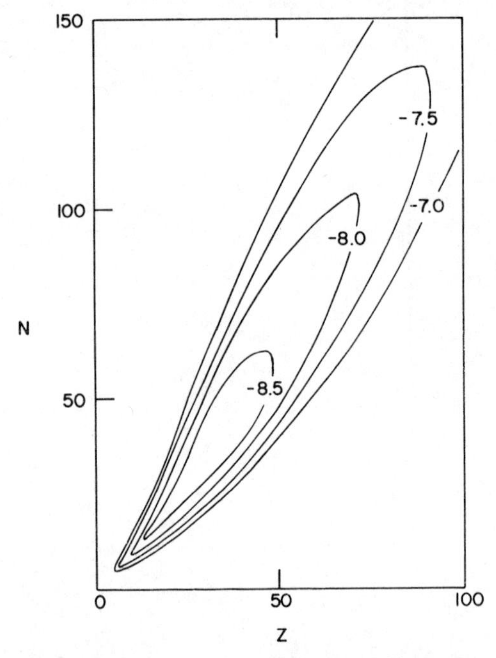

Figure 14.5 Contour map of the 'valley of stability' (actually a long, narrow trough).

Pairing energy. Our semi-empirical mass formula (14.12) predicts, of course, the same parabolic variation of e for even-A isobars as for odd-A isobars. However, the situation is complicated by an effect that we have not hitherto considered, that of the tendency for nucleons of the same kind to pair off. By this we mean that for any nucleus in its ground state there is an additional binding energy $\frac{1}{2}\Delta$, over and above what is predicted by

(14.12), associated with an even number of nucleons of the same kind, and an equal loss of binding associated with an odd number of nucleons of the same kind.

The net effect of this will, of course, be zero in the case of odd-A nuclei, for which there are an even number of one kind, and an odd number of the other kind of nucleon. But nuclei with an even number of both protons and neutrons ('even–even' nuclei) will have their total binding energies enhanced by Δ with respect to (14.12)†, while nuclei with an odd number of each kind of nucleon ('odd–odd' nuclei) will lose binding energy Δ. Thus we should write in place of (14.12) as our final form of the semi-empirical mass formula

$$e = -a_{\text{vol}} + a_{\text{surf}}\frac{1}{A^{1/3}} + a_C\frac{Z^2}{A^{4/3}} + a_{\text{sym}}\left(\frac{N-Z}{A}\right)^2 + \frac{\delta}{A} \tag{14.31}$$

where

$$\delta = \begin{cases} -\Delta & Z \text{ even, } N \text{ even} \\ 0 & A \text{ odd} \\ \Delta & Z \text{ odd, } N \text{ odd} \end{cases} \tag{14.32}$$

in which it is found empirically that the so-called *pairing energy* can be represented by

$$\Delta \simeq \frac{11}{A^{1/2}}\ \text{MeV}. \tag{14.33}$$

Now in the case of even A, we have alternately even–even and odd–odd nuclei as Z varies. Since the former will be depressed below the mass parabola of figure 14.4 and the latter raised above, it follows that for even A we shall have *two* parabolas, separated by distance $2\Delta/A$ (see figure 14.6).

In view of this pairing effect it is easy to understand why well over half of the 280 stable nuclides are even–even, while only four are odd–odd (see §16).

Shell corrections. Another microscopic effect concerns the slight variations of binding energy that are associated with nuclear shell structure; this is in fact one of the principal manifestations of the shell model (§12). We shall not consider these corrections here, but it should be noted that they must be taken into account if we are to have a precision determination of the coefficients that we do consider explicitly, i.e. those appearing in equation (14.31).

† This effect will be operative only between pairs of nucleons whose total angular momentum is zero. This is why even–even nuclei invariably have zero total angular momentum J for their ground states.

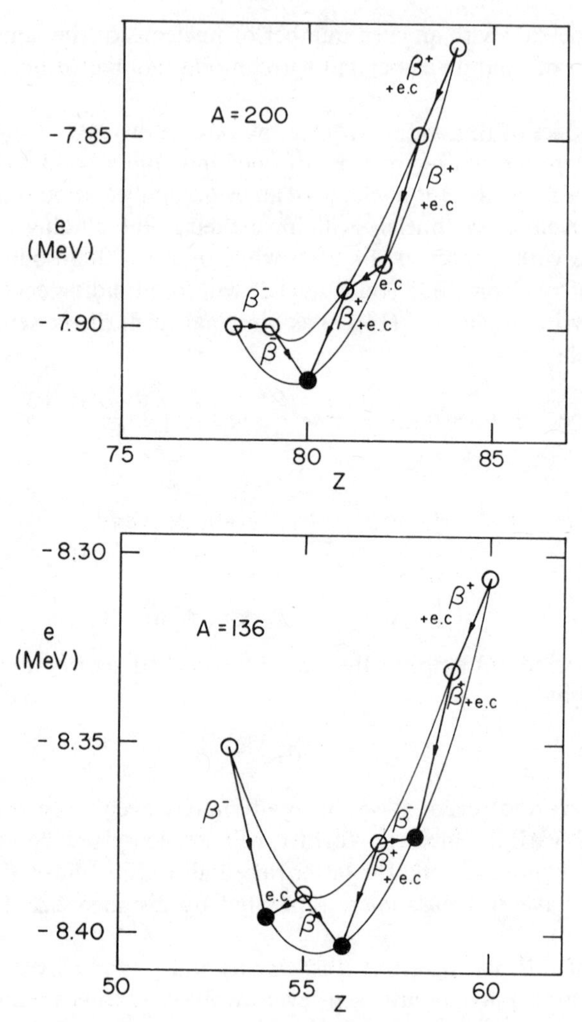

Figure 14.6 Isobaric variation of binding energy of even-A nuclei. The arrows (β transitions) are explained in §16; the full circles represent the β stable nuclei.

Theory of nuclear matter. We have remarked on the attempts that have been made to understand nuclear properties in terms of the basic nucleon–nucleon force, and on the very great complexity of such calculations. Now by far the simplest nuclear system with more than two nucleons is infinite nuclear matter, since its uniformity and absence of a surface means that it must be translationally invariant. Of course, we have seen that such a system could exist only in the absence of the Coulomb force, and thus must be hypothetical. Nevertheless, we know what its binding energy per nucleon is: it is simply the volume term of the semi-empirical mass formula, if the densities of neutrons and protons are equal. Likewise, its density can be inferred from that of real nuclei (§4).

Thus it is natural that before attempting to calculate the properties of real nuclei one first considers the case of infinite nuclear matter, since it can effectively serve as a 'test-bed' for studying both the validity of various nucleon–nucleon forces, and also of the calculational techniques. An enormous effort has in fact been devoted to this problem, but at the time of writing it has still not been completely resolved. This explains why attempts to calculate real, finite nuclei from the basic force have been even less successful. For the same reason, much less effort has been expended on the attempt to derive the other parameters of the semi-empirical formula from the nucleon–nucleon force.

§15 THE ROLE OF GRAVITY

We have seen in the previous section how the joint action of the Coulomb and symmetry terms lead to an apparent upper limit of between 250 and 300 on the mass numbers of nuclei that can exist. More particularly, if the Coulomb force could be switched off there would be no such upper limit and we could have nuclei with indefinitely large numbers of nucleons. The increasing loss of binding that the Coulomb force engenders as A increases is due, we recall, to the fact that this force is both repulsive, and, being long-range, non-saturating.

Now there exists in nature another force that is non-saturating, namely the gravitational force, which is, in fact, formally identical to the Coulomb force, except that it is attractive. Since it will certainly act between the nucleons of a nucleus (between all pairs of nucleons, in fact, and not just between protons) it will tend to have the opposite effect to the Coulomb force, i.e. it will tend to make the binding energy *increase* with mass number rather than decrease. However, the gravitational force is so much weaker than the Coulomb force that it is utterly negligible in all ordinary nuclei, and our failure to consider it so far is quite justified. Nevertheless, we shall now examine its implications for the semi-empirical mass formula.

In analogy with the electrostatics result (14.7), the gravitational energy of a uniform sphere of mass $\mathcal{M}$ and radius R is

$$\Omega = -\frac{3}{5}\frac{G\mathcal{M}^2}{R} \tag{15.1}$$

where G is the gravitational constant ($=6.67\times10^{-11}\,\mathrm{m^3\,kg^{-1}\,s^{-2}}$). Then since we can write $\mathcal{M}=MA$ and $R=r_0A^{1/3}$ for a nucleus, the gravitational energy can be incorporated with the Coulomb term, so that the semi-empirical formula will read simply

$$e=-a_{\mathrm{vol}}+a_{\mathrm{surf}}\frac{1}{A^{1/3}}+\frac{3}{5r_0}(\gamma^2e^2-GM^2)A^{2/3}+a_{\mathrm{sym}}(1-2\gamma)^2 \tag{15.2}$$

where $\gamma=Z/A$, the fraction of nucleons that are protons, and we have neglected the pairing term.

Now

$$\frac{GM^2}{e^2}=8.07\times10^{-37} \tag{15.3}$$

so that the gravitational force is so weak compared with the electrostatic force that it will be utterly negligible unless γ is very close to zero, i.e.unless the nucleus consists almost entirely of neutrons. Putting $\gamma=0$, then, we have

$$e=-a_{\mathrm{vol}}+a_{\mathrm{surf}}\frac{1}{A^{1/3}}+a_{\mathrm{sym}}-\frac{3GM^2}{5r_0}A^{2/3}. \tag{15.4}$$

Now $a_{\mathrm{sym}}-a_{\mathrm{vol}}\simeq 15$ MeV†, and it is easy to verify that compared with this the gravitational term will be quite negligible for 'ordinary' values of A, even when $\gamma=0$. Thus e will be positive and we shall have confirmed the common statement that nuclei consisting entirely of neutrons are unbound.

However, the gravitational term, being non-saturating, becomes increasingly important as A increases, while the other terms remain constant, or decrease. Furthermore, the gravitational force is attractive, so we see that for large enough A a nucleus composed entirely of neutrons can indeed be bound, although it will be the gravitational rather than the nuclear forces that are responsible. Neglecting the surface term, we see that this will happen for

$$\frac{3}{5r_0}GM^2A^{2/3}\geqslant a_{\mathrm{sym}}-a_{\mathrm{vol}} \tag{15.5}$$

i.e. $A\geqslant10^{56}$, which corresponds to a 'nucleus' with a mass of at least 10^{29} kg. Since this critical mass is of the order of one tenth of the mass of the sun we see that when the semi-empirical mass formula is generalised to include gravity it effectively predicts the existence of stars that consist essentially of a single giant nucleus, composed entirely of neutrons. The existence of such *neutron stars* had been conjectured since the discovery of the neutron itself in 1932, and was finally confirmed in 1968 with the discovery of the so-called *pulsars*, stars that are apparently rotating with a

† We are using here the refined values given in (14.19).

period of the order of a second or even less. That these objects could indeed be identified as the giant nuclei predicted by the semi-empirical mass formula (15.4) followed from the fact that although their masses were of the order of a solar mass, their radii were only around five or ten kilometres, which is just what is given by the usual nuclear-radius relationship (4.2) for a mass number of $A \simeq 10^{56}$, so that the density is about the same as for an ordinary nucleus, i.e. several hundred million tonnes per cubic centimetre.

In fact, this constancy of the nuclear density was tacitly supposed in the form (15.4) of the mass formula when we assumed that even for such enormous mass numbers we could use the same values of the coefficients as for ordinary nuclei. However, we show in Appendix D that when the binding is primarily gravitational increasing mass leads to such a large increase in density that the radius will actually decrease, essentially because of the non-saturating character of the gravitational force†. The paradox is resolved by noting that the masses of the neutron stars fall within a very narrow range of between one and two solar masses, for which it turns out that the corresponding density is just about that of ordinary nuclei. Heavier neutron stars, whose density would be supra-nuclear, are unstable, collapsing to form the so-called 'black holes', while lighter objects do not form neutron stars at all, but instead stabilise as white dwarfs, as explained in Appendix D.

The reader will have begun to appreciate by now that our attempt to describe neutron stars in terms of the straightforward gravitational generalisation of the semi-empirical mass formula constitutes a gross over-simplification. It is remarkable (and possibly fortuitous) that the estimate given by (15.5) for the lower limit on the mass of a neutron star is in error by no more than one order of magnitude. In fact, there are several respects in which the interpretation of a neutron star simply as an overgrown version of an ordinary nucleus breaks down. For example, a neutron star, unlike an ordinary nucleus, will contain a slight admixture (about 3% by number) of electrons, along with an equal number of protons, the whole remaining electrically neutral‡. These electron–proton pairs may be regarded as arising from the β decay of the neutron, which continues until the Fermi energy of the resulting electron gas exceeds 782 keV, the energy available from the neutron decay (see §16). Furthermore, our interpretation of a neutron star as a single gigantic nucleus is marred by the fact that towards the surface

†See equation (D.23). Actually, the analysis of Appendix D ignores completely the contribution of the nuclear forces to the binding and thus will exaggerate the increase of density with the mass of the neutron star. Nevertheless, the point remains that we have no right to assume that the density is independent of the mass.

‡To distinguish this complicated mixture which exists in real neutron stars from the much simpler but hypothetical nuclear matter of theorists, we may call it 'celestial nuclear matter'.

the gas consisting of neutrons, protons and electrons breaks up into small clusters that resemble ordinary nuclei more and more closely, with the crust itself composed of more or less conventional atoms.

PROBLEMS FOR CHAPTER II

II.1 Show that the correctly antisymmetrised wavefunction of the independent-particle model can be expressed as the so-called Slater determinant

$$\Psi_0(\boldsymbol{r}_1, \ldots, \boldsymbol{r}_A) = \frac{1}{\sqrt{A!}} \begin{vmatrix} \phi_1(1) & \phi_1(2) & \ldots & \phi_1(A) \\ \phi_2(1) & \phi_2(2) & \ldots & \phi_2(A) \\ \vdots & & & \\ \phi_A(1) & \phi_A(2) & \ldots & \phi_A(A) \end{vmatrix}$$

where ϕ denotes a single-particle function of equation (12.10).

II.2 Consider a spherical nucleus with uniform charge density, radius R, i.e.

$$\rho(r) = \begin{cases} \rho_0 & r < R \\ 0 & r > R. \end{cases}$$

(This is just the Fermi distribution (4.1) with $a = 0$.) Show that the Coulomb potential in which the individual protons move is given by

$$V_C(r) = \begin{cases} \dfrac{Ze^2}{2R}\left(3 - \dfrac{r^2}{R^2}\right) & r < R \\ \\ \dfrac{Ze^2}{r} & r > R \end{cases}$$

(This is the term that appears in equation (12.22) for the complete shell-model potential.)

II.3 Referring to figure 12.1 for the shell-model level scheme, assume the following values for some of the single-particle energies:

$1f_{7/2}$	-28.6 MeV
$1f_{5/2}$	-25.9
$2p_{3/2}$	-22.7
$2p_{1/2}$	-21.7
$1g_{9/2}$	-19.2
$1g_{7/2}$	-15.0
$2d_{5/2}$	-12.2
$2d_{3/2}$	-10.3
$1h_{11/2}$	-9.5
$3s_{1/2}$	-9.3.

(These values come from a Hartree-Fock calculation on ^{208}Pb.) Verify to what extent the spin–orbit splitting is proportional to $(2l+1)$, as indicated by equation (12.28). Explain the deviations.

II.4 If the sign of the $\boldsymbol{s}\cdot\boldsymbol{l}$ term in the shell-model potential (12.22) is reversed, determine the new magic numbers.

II.5 In metallic copper (density $9\,\mathrm{g\,cm^{-3}}$, atomic weight 63.5) there is about one conduction electron per atom. Show that the electron gas is degenerate at room temperature. Show that a neutron gas with the same density as copper will not be degenerate at this temperature.

II.6 Prove equation (14.6), referring to figure 14.3 and making any reasonable assumptions.

II.7 Prove equation (14.7) (make use of the results of Problem II.2).

II.8 Calculate the coefficient a_C of the mass formula (14.12) from the masses of all pairs of mirror nuclei with odd A that you can find in the mass table in Appendix G.

II.9 Verify equation (14.16) for the energy of nuclei on the stability line, and show that the last term therein can be rewritten according to

$$\frac{a_C}{2x+4A^{-2/3}} = a_C\,\frac{Z_0(A)}{2A^{1/3}}\,.$$

II.10 Use the mass formula (14.12) to calculate the largest value of A for which nuclei on the stability line can be bound.

II.11 Add to the mass formula (14.12) an extra volume-symmetry term $b_{\mathrm{sym}}I^4$. Calculate the coefficient b_{sym} in the same way that the coefficient a_{sym} of the term in I^2 was calculated on pp. 78–80.

II.12 Using equations (14.29) and (14.30), determine the value of a_{sym} implied by each of the two curves in figure 14.4.

CHAPTER III

NUCLEAR INSTABILITY

We pointed out in §10 that nuclei can transform spontaneously into other nuclei of greater binding energy. It will be recalled that there are two main categories of such transformations: β transformations and fragmentation. β transformation is an isobaric process in which a neutron of the nucleus in question changes into a proton, or vice versa. In the fragmentation process the nucleus in question breaks up into two smaller nuclei, the total number of neutrons and the total number of protons each being conserved.

In the present chapter we examine the conditions that determine whether or not a given nucleus is unstable with respect to these different modes of transformation; we shall make particular use of the binding-energy systematics established in §14. We deal first with β transformations (§16), and then with fragmentation (§17).

Having established that a particular nucleus is unstable with respect to a particular mode of transformation, there arises the question as to the rate of the corresponding decay, i.e. the half-life, as defined in §10. We have discussed this question in the context of β decay in §11: it will be recalled that ultimately β decay rates depend on the strength of the weak interaction, but that there are wide variations from one nucleus to another. As for the rates of fragmentation processes, these depend to a large extent on the Coulomb barrier which the separating charged nuclei must penetrate. This question will be discussed in §18; we shall see there that many energetically possible fragmentation processes are simply not observed, so great is the inhibiting effect of the Coulomb barrier.

§16 THE ENERGETICS OF β TRANSFORMATIONS

We begin by noting that the total energy available for a β transition is given simply by the difference in mass between the initial and final nuclei

$$Q = (M_{\text{nuc}}(Z, N) - M_{\text{nuc}}(Z \pm 1, N \mp 1))c^2 \qquad (16.1)$$

where the upper sign refers to β^- decay and the lower to β^+ decay (and electron capture). Writing this in terms of the internal energy per nucleon, we find

$$Q = A(e(Z, N) - e(Z \pm 1, N \mp 1)) \pm (M_n - M_p)c^2. \qquad (16.2)$$

On the other hand, expressing the nuclear masses in terms of atomic masses, we have from equation (5.7)

$$Q = (M_{\text{at}}(Z, N) - M_{\text{at}}(Z \pm 1, N \mp 1))c^2 \pm m_e c^2 \qquad (16.3)$$

which in terms of the mass excess (5.10) is

$$Q = \Delta(Z, N) - \Delta(Z \pm 1, N \mp 1) \pm m_e c^2. \qquad (16.4)$$

In fact, because of the binding energy of the atomic electrons, the value of Q given by equations (16.1) or (16.2) will not be identical to that given by equations (16.3) or (16.4). Denoting the former by Q_{nuc} and the latter by Q_{at} we shall have, using equation (5.9)

$$Q_{\text{at}} - Q_{\text{nuc}} = \pm 34.25\, Z^{1.39} \text{ eV}. \qquad (16.5)$$

Usually this difference is quite negligible, but for certain nuclei it can be critical in determining whether or not we have stability (see below). Usually we shall ignore the difference.

We now examine the conditions for stability of a nucleus against β transformations. For $\beta^{\pm}$ decay an electron has to be created, so assuming that the neutrino mass is zero, and that the recoil energy of the final (daughter) nucleus is negligible (see Problem III.1), we find that the maximum kinetic energy of the β electron (or of the neutrino, for that matter) is

$$T_0 = Q - m_e c^2 \qquad (16.6)$$

so that the condition for β decay to occur is

$$Q > m_e c^2. \qquad (16.7)$$

On the other hand capture of orbital electrons involves the *destruction* of an electron, so the emitted neutrino will have a unique energy

$$E_\nu = Q + m_e c^2 \qquad (16.8)$$

where we have neglected the binding energy of the captured electron in its atomic orbit. Thus the condition for electron capture to occur is

$$Q > -m_e c^2 \tag{16.9}$$

which is less stringent than the condition for β^+ decay. We see that if the Q value of a candidate for a β^- decay is less than the $m_e c^2$ required by (16.7) for the creation of the electron, then, even if it is positive, not only will the β^- decay not take place but we shall have an electron capture in the opposite direction.

Let us next rewrite these results in terms of the expression (16.2) for Q. For β^- decay (16.6) becomes

$$T_0 = A(e(Z,N) - e(Z+1, N-1)) + T_{0n} \tag{16.10}$$

where

$$T_{0n} = (M_n - M_p - m_e)c^2 = 782 \text{ keV}. \tag{16.11}$$

This latter quantity will clearly be equal to the maximum kinetic energy of the electron in the β decay of the free neutron, since both $e(Z,N)$ and $e(Z+1, N-1)$ are zero in that case. For β^+ decay, on the other hand, we shall have

$$T_0 = A(e(Z,N) - e(Z-1, N+1)) - T_{0n} - 2m_e c^2 \tag{16.12}$$

while for electron capture (16.8) gives

$$E_\nu = A(e(Z,N) - e(Z-1, N+1)) - T_{0n}. \tag{16.13}$$

The energy conditions for the various processes to occur are then as follows

β^- decay $\quad A(e(Z,N) - e(Z+1, N-1)) > -T_{0n}$ (16.14a)

β^+ decay $\quad A(e(Z,N) - e(Z-1, N+1)) > T_{0n} + 2m_e c^2$ (16.14b)

electron capture $\quad A(e(Z,N) - e(Z-1, N+1)) > T_{0n}$. (16.14c)

Because of the non-zero value of T_{0n}, i.e. because of the instability of the free neutron, it follows that the lowermost nucleus on the internal-energy parabola is not necessarily stable: there can actually be an 'uphill' β^- decay, as will be seen in figures 14.4 and 14.6. However, the effect is not marked and never involves a shift of more than unity in the Z value of the most stable nucleus, and then it is from one side of the minimum to the other (see Problem III.2)†.

†The effect could have been eliminated altogether if we had plotted nuclear masses rather than internal energies in figures 14.4 and 14.6, as do many authors. Our own choice appears to be the more convenient for most purposes. The Z_0 values corresponding to the minima in the two representations never differ by more than unity.

Expressing Q in terms of atomic masses, as in (16.3), we have for β^- decay

$$T_0 = (M_{\rm at}(Z, N) - M_{\rm at}(Z+1, N-1))c^2 \tag{16.15}$$

and for β^+ decay

$$T_0 = (M_{\rm at}(Z, N) - M_{\rm at}(Z-1, N+1))c^2 - 2m_e c^2 \tag{16.16}$$

while for electron capture

$$E_\nu = (M_{\rm at}(Z, N) - M_{\rm at}(Z-1, N+1))c^2. \tag{16.17}$$

The presence of the term $2m_e c^2$ in equation (16.16), but not in equations (16.15) or (16.17), can be understood by considering the decay process as one involving the entire neutral atom: in β^+ decay there has to be creation of an electron–positron pair, but for the other two processes there is no net creation or destruction (except of massless neutrinos). It is easy to rewrite equations (16.15), (16.16) and (16.17) in terms of the mass excess (5.10).

The instability conditions are now

β^- decay	$M_{\rm at}(Z, N) > M_{\rm at}(Z+1, N-1)$	(16.18a)
β^+ decay	$M_{\rm at}(Z, N) > M_{\rm at}(Z-1, N+1) + 2m_e$	(16.18b)
electron capture	$M_{\rm at}(Z, N) > M_{\rm at}(Z-1, N+1)$.	(16.18c)

This form of the instability conditions is probably clearer than the form (16.14), and a number of conclusions immediately follow.

Firstly, we see that for a nucleus (Z, N) to be stable with respect to all β transitions it is necessary and sufficient that both adjacent isobars $(Z \pm 1, N \mp 1)$ have a greater atomic mass. By the same token it follows that both the immediate neighbours of a β stable nucleus will themselves suffer from some form of β instability, i.e. if (Z, N) is β stable then both $(Z \pm 1, N \mp 1)$ must be β unstable. Thus all pairs of adjacent isobars will be connected by some β transformation, as shown by the arrows in figures 14.4 and 14.6: if the transition does not take place in one direction it will take place in the other.

We see from these two figures that stable nuclei will have two arrows leading into them, i.e. they are fed from both sides. Clearly, in the case of odd-A nuclei, with their single parabola, there can be only one stable nucleus, and no nucleus can decay in both directions, i.e. undergo both β^+ decay (or electron capture) and β^- decay. The situation is considerably more complicated in the case of even-A nuclei, and neither of these restrictions holds: in figure 14.6 it will be seen that decay in both directions is possible and that there can be more than one stable nucleus. Usually these stable nuclei lie on the lower parabola, i.e. are even–even, but for $A = 6$, 10 and 14 the stable nuclei are odd–odd (^{6}Li, ^{10}B and ^{14}N respectively), although there is only one for each A value.

Bare nuclei. So far, in discussing this question of stability, we have not distinguished between Q_{at} and Q_{nuc} (see equation (16.5) and the preceding discussion). It is conceivable that there will be cases where neutral atoms will be able to undergo β^- decay, but the corresponding bare nuclei will not (see Problem III.6).

Neutronisation in collapsing stars. We now consider the energetics of the electron capture that takes place during the formation of neutron stars (see §§11 and 28 and Appendix D). For simplicity we suppose that the protons of the star are free, although in reality they are all bound in relatively complex nuclei by the time the star is beginning to collapse. The necessary electron energy for the reaction (11.15)

$$p + e^- \rightarrow n + \nu \tag{16.19}$$

to proceed is just

$$T_e = (M_n - M_p - m_e)c^2 \equiv T_{0n} \tag{16.20}$$

i.e. 782 keV.

As the density of the collapsing star increases its electrons become degenerate and the Fermi energy rises according to equation (C.11). Actually, this equation is non-relativistic, and for electrons we should replace it by the easily proven relation

$$n = \frac{1}{3\pi^2}\left(\frac{m_e c}{\hbar}\right)^3\left[\left(1 + \frac{\varepsilon_F}{m_e c^2}\right)^2 - 1\right]^{3/2} \tag{16.21}$$

which is valid for all energies. Once the Fermi energy reaches T_{0n}, which it will do at a density of $n_0 = 7.37 \times 10^{30}$ cm^{-3}, the reaction (16.19) can begin, and all electrons whose energy exceeds T_{0n} will be captured by protons to form neutrons.

Thus as the star continues to contract electrons will be removed by the process (16.19) at such a rate that their density remains constant at n_0; the density of the protons will likewise remain constant at this same value, of course. Clearly, the contracting star becomes progressively richer in neutrons. Because any excess energy available in the reaction (16.19) will be immediately carried out of the star by the neutrinos, which interact only negligibly with the stellar material (at this stage), the neutrons thus formed will, if they 'attempt' to β decay back to protons, be unable to emit electrons with energy greater than the usual 782 keV. But such β decay will be blocked by the operation of the Pauli principle, all available electron states being already occupied, so that the neutrons are effectively stable. Of course, if ever the star began to expand again, β decay of the neutron would resume.

It should be realised that throughout this discussion we have been supposing that the neutron and proton gases have sufficiently low density to be completely non-degenerate. Passing to the opposite extreme of completely degenerate nucleon gases it is easy to see that we must have

$$\varepsilon_{\mathrm{Fe}} + \varepsilon_{\mathrm{Fp}} - \varepsilon_{\mathrm{Fn}} = T_{0\mathrm{n}} \tag{16.22}$$

where $\varepsilon_{\mathrm{Fe}}$, $\varepsilon_{\mathrm{Fp}}$, and $\varepsilon_{\mathrm{Fn}}$ are the Fermi energies of the electron, proton and neutron gases respectively.

Capture of neutrinos. We conclude this section by considering the capture of neutrinos by nuclei that are otherwise stable, as in equations (11.17) and (11.18)

$$(Z, N) + \nu \rightarrow (Z+1, N-1) + \mathrm{e}^- \tag{16.23a}$$

$$(Z, N) + \bar{\nu} \rightarrow (Z-1, N+1) + \mathrm{e}^+ . \tag{16.23b}$$

The threshold energies of the neutrinos for these processes to occur are given by

$$\begin{aligned} E_\nu^{\min} &= (M_{\mathrm{nuc}}(Z+1, N-1) - M_{\mathrm{nuc}}(Z, N) + m_{\mathrm{e}})c^2 \\ &= (M_{\mathrm{at}}(Z+1, N-1) - M_{\mathrm{at}}(Z, N))c^2 \\ &= \Delta(Z+1, N-1) - \Delta(Z, N) \end{aligned} \tag{16.24a}$$

and

$$\begin{aligned} E_{\bar{\nu}}^{\min} &= (M_{\mathrm{nuc}}(Z-1, N+1) - M_{\mathrm{nuc}}(Z, N) + m_{\mathrm{e}})c^2 \\ &= (M_{\mathrm{at}}(Z-1, N+1) - M_{\mathrm{at}}(Z, N) + 2m_{\mathrm{e}})c^2 \\ &= \Delta(Z-1, N+1) - \Delta(Z, N) + 2m_{\mathrm{e}}c^2 \end{aligned} \tag{16.24b}$$

respectively.

§17 FRAGMENTATION: ENERGETICS

A nucleus (Z, N), i.e. a nucleus having Z protons and N neutrons, will be energetically capable of *spontaneously* disintegrating into two fragments, (Z', N') and $(Z - Z', N - N')$, respectively, if the atomic masses involved satisfy the inequality

$$M(Z,N) \geqslant M(Z',N') + M(Z-Z',N-N'). \tag{17.1}$$

Since the same numbers of neutrons and protons are involved on both sides of (17.1) we can rewrite it in terms of the mass excess, defined in equation (5.10), as

$$\Delta(Z,N) \geqslant \Delta(Z',N') + \Delta(Z-Z',N-N') \tag{17.2}$$

and in terms of the internal energy per nucleon as

$$Ae(Z,N) \geqslant A'e(Z',N') + (A-A')e(Z-Z',N-N') \tag{17.3}$$

where we are assuming that the *change* in the total electron binding energy during the disintegration is negligible†.

When the instability condition (17.1) is satisfied the energy released in the corresponding process will be

$$\begin{aligned} Q &= (M(Z,N) - M(Z-Z',N-N') - M(Z',N'))c^2 \\ &= \Delta(Z,N) - \Delta(Z-Z',N-N') - \Delta(Z',N') \\ &= Ae(Z,N) - (A-A')e(Z-Z',N-N') - A'e(Z',N'). \end{aligned} \tag{17.4}$$

This energy will be carried off as kinetic energy of the two fragments; with momentum being conserved we shall have for the kinetic energy of the fragment (Z',N')

$$K(Z',N') = \left(1 - \frac{A'}{A}\right)Q \tag{17.5}$$

where in the kinematical calculation we have taken the masses to be rigorously proportional to the mass numbers.

When (17.1) is not satisfied, so that our nucleus (Z,N) is stable with respect to the indicated fragmentation, the energy Q will be negative. It is then customary to speak of the *separation energy*, $S = -Q$, which is simply the energy that must be supplied in order for the fragmentation to take place.

Nucleon emission. We consider now the simplest possible form of disintegration, the emission of a single nucleon. Since the corresponding binding energy is obviously zero, equation (17.3) becomes

$$Ae(A) \geqslant (A-1)e(A-1) \tag{17.6}$$

where $e(A) \equiv e(Z,N)$ and $e(A-1) \equiv e(Z,N-1)$ or $e(Z-1,N)$, according to whether we are dealing with neutron or proton emission, respectively.

†For this reason we shall work exclusively in terms of *atomic* masses in this section.

If we write now $e(A-1) \equiv e$ and $e(A) \equiv e + \Delta e$, we get finally as the condition for single-nucleon emission

$$\Delta e \geqslant -\frac{e}{A}. \tag{17.7}$$

Since $-e/A$ is positive it follows that along the stability line it will only be for $A > 56$ that there will be any chance of satisfying (17.7), since $\mathrm{d}e/\mathrm{d}A$ is negative for $A < 56$. For $A > 56$ we can use figure 14.2 to write approximately

$$e(A) = -9.13 + 0.0068A \text{ MeV} \qquad A > 56 \tag{17.8}$$

from which $\Delta e = 0.0068$ MeV. Then using (17.8) again, we find that the inequality (17.7) can only be satisfied for $A > 670$, and we predict that no real nucleus lying on the line of β stability can ever emit nucleons from its ground state.

In fact, this argument, based on a simplified semi-empirical formula, overlooks the considerable fluctuations that individual masses exhibit, mainly because of shell-model effects, and it turns out that for $A = 5$ there is no nucleus that is stable with respect to nucleon emission. Indeed, if ever an $A = 5$ system is formed the subsequent nucleon emission is so rapid that we can say that $A = 5$ nuclei just do not exist (see §28 for the astrophysical significance of this).

However, if we move far enough away from the stability line nucleon emission becomes the rule, and no longer depends on abnormally large shell-model effects. To see this, we use the form (14.29) of the nuclear internal energy, so that with (14.30) the instability condition (17.6) becomes, neglecting pairing

$$Ae_0(A) - (A-1)e_0(A-1) + 2a_{\text{sym}}\left(\frac{(Z-Z_0(A))^2}{Z_0(A)} - \frac{(Z-Z_0(A-1))^2}{Z_0(A-1)}\right) > 0 \tag{17.9a}$$

for neutron emission, and

$$Ae_0(A) - (A-1)e_0(A-1) + 2a_{\text{sym}}\left(\frac{(Z-Z_0(A))^2}{Z_0(A)} - \frac{(Z-1-Z_0(A-1))^2}{Z_0(A-1)}\right) > 0 \tag{17.9b}$$

for proton emission. Now from equation (14.13) we find, after a little algebra (Problem III.9)

$$Z_0(A-1) \simeq Z_0(A) - \tfrac{1}{2} + \varepsilon \tag{17.10}$$

provided that ε, given by

$$\varepsilon = \frac{5x}{12}A^{2/3} \tag{17.11}$$

is sufficiently small. Then (17.9) becomes, writing $Z_0(A) \equiv Z_0$

$$Ae_0(A) - (A-1)e_0(A-1) + 2a_{\text{sym}}\left(\frac{(Z-Z_0)^2}{Z_0} - \frac{(Z-Z_0-\varepsilon \pm \frac{1}{2})^2}{Z_0 - \frac{1}{2} + \varepsilon}\right) > 0 \qquad (17.12)$$

where the upper sign refers to neutron emission and the lower to proton emission. If we assume now that the internal energy per nucleon is roughly constant along the stability line, i.e. that $e_0(A) = e_0(A-1) \equiv e_0$, and also that $Z_0 \gg |Z - Z_0| \gg 1$, then with $\varepsilon \ll 1$, (17.12) simplifies to

$$e_0 + \frac{2a_{\text{sym}}}{Z_0}(2\varepsilon \mp 1)(Z - Z_0) > 0. \qquad (17.13)$$

This means that we shall have instability with respect to neutron emission for neutron-rich nuclei, i.e. proton-deficient nuclei, such that

$$Z < Z_0\left(1 + \frac{e_0}{2a_{\text{sym}}}\frac{1}{1-2\varepsilon}\right) \simeq Z_0\left(1 - \frac{0.2}{1-2\varepsilon}\right) \qquad (17.14a)$$

and instability with respect to proton emission for proton-rich nuclei such that

$$Z > Z_0\left(1 - \frac{e_0}{2a_{\text{sym}}}\frac{1}{1+2\varepsilon}\right) \simeq Z_0\left(1 + \frac{0.2}{1+2\varepsilon}\right). \qquad (17.14b)$$

These limits define the so-called *nucleon drip lines* shown in figure 17.1. Nuclei lying below the neutron drip line satisfy (17.14a) and therefore tend to spontaneously lose neutrons, while those lying above the proton drip line satisfy (17.14b) and therefore tend to emit protons†. The neutron drip line is of special significance in connection with the electron-capture process that takes place during the formation of neutron stars, as discussed in the previous section. The electron-capturing protons are actually bound in nuclei, so that as capture proceeds the nuclei drift away from the stability line, becoming richer and richer in neutrons. Finally, when the nuclei reach

†The approximations made in going from equation (17.9) to equation (17.14) lead to significant errors for the neutron drip line when $A > 160$. The drip lines shown in figure 17.1 have been drawn, therefore, to correspond to equation (17.9), rather than equation (17.14).

the neutron drip line neutron emission begins, and the nuclei start to dissolve into a neutron gas.

We now note that it should also be possible for nuclei lying between the drip lines to emit nucleons. In the first place, because of the fluctuations that individual nuclear masses exhibit about the smooth trends of our semi-empirical formula, there should be occasional nuclei between the drip lines that are nucleon-unstable: we have already referred to one such case existing on the stability line itself ($A = 5$ nuclei). Furthermore, it should be particularly emphasised that throughout this entire discussion of nucleon radioactivity we have been supposing that the nuclei in question have been in their ground states. It is clear now that the condition (17.7) for nucleon emission can in principle be satisfied for any nucleus provided that it is sufficiently highly excited.

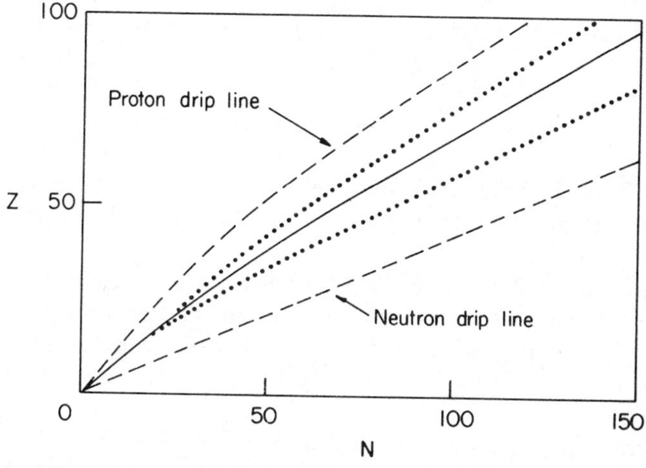

Figure 17.1 The full curve represents the line of β stability, as in figure 14.1. Outside the drip lines (broken), defined by equations (17.9), there is instability with respect to nucleon emission. Between each drip line and the adjacent dotted line, defined by equations (17.21), there is instability with respect to delayed nucleon emission.

Nevertheless, there remains one serious obstacle to the observation, or rather the identification, of nucleon radioactivity, and that is the fact that usually the nucleon emission takes place extremely rapidly, a very short time indeed after the formation of the nucleon-unstable nucleus (this is why we can say, for example, that the $A = 5$ nuclei are non-existent). This is in contrast to what happens in the case of α decay, and is related to the phenomenon of barrier penetration (see the next section). For neutron emission there is no Coulomb barrier at all, so that the emission is effectively instantaneous, while for proton emission there is a barrier but it is very weak (since $Z_1 = 1$ in (18.6) rather than 2, as in the case of α decay), so that the emission is very rapid, unless the available energy is exceptionally small.

To understand the difficulty which arises because of the rapidity of the nucleon emission, consider first the case of proton emission. Now the obvious way to form proton-rich nuclei is by means of nuclear reactions, i.e. bombarding one nucleus with another (see Chapter IV). For example, if we bombard ^{120}Sn with ^{40}Ca, both of which lie on the stability line, and if the two simply stick together, then we shall form ^{160}Yb

$$^{40}Ca + {}^{120}Sn \rightarrow {}^{160}Yb \tag{17.15}$$

which lies far from the stability line on the proton-rich side. (If at the same time some neutrons are released, as often happens, then the product nucleus will be still more proton rich.) However, because of the rapidity with which any proton emission would take place, it would be difficult to attribute it unequivocally to the formation of a proton-emitting nucleus, rather than to a fragmentation process occurring during the reaction itself.

As for neutron-rich nuclei, these are most readily formed as fission products. It is found experimentally that fission is indeed accompanied by the emission of several neutrons, but insofar as this emission is extremely rapid it is again difficult to say whether the neutrons are released as an essential part of the fission process, or whether truly neutron-emitting nuclei are formed as fission products.

Actually, it is possible in principle to settle the question by measuring the angular correlation between the nucleon emission and the final recoiling nucleus. However, even when it is clear that we really do have nucleon-emitting nuclei the process is usually too rapid for any timescale to be measurable.

The first case of true nucleon radioactivity, in which the emission follows the decay law (10.1) with a measurable half-life, was found only in 1970: this was the proton decay of an excited state of ^{53}Co. Now this nucleus is easily seen to lie inside the proton drip line, and the ground state is indeed found to be stable against proton decay. It was not until 1982 that nucleon radioactivity was observed in a nuclear ground state: this was the proton radioactivity of ^{151}Lu, with a half-life of 85 ms. (This nucleus lies just inside the proton drip line of figure 17.1, but it should be realised that equation (17.9), on which the drip lines of this figure are based, neglects completely all shell effects, and even the smooth trends are only represented within the limitations of the Bethe–von Weizsäcker mass formula (14.12).)

Delayed nucleon emission. The very rare phenomenon of true nucleon radioactivity should not be confused with what is known as *delayed nucleon emission*, in which the neutron-rich fission product, or the proton-rich reaction product, first undergoes a β decay (electron or positron, respectively) before emitting a nucleon (neutron or proton, respectively). The nucleon emission may still occur effectively simultaneously with the formation of the nucleon-unstable daughter product of the β decay, but will be delayed by the finite timescale of the β decay itself. The nucleon emission will in fact be observed to obey the radioactive decay law (10.1), with a mean lifetime equal to that of the preceding β decay (see equation (10.12)).

Many cases of such delayed neutron and delayed proton emission are known. For example, although most of the fission neutrons are released extremely rapidly, and are thus known as *prompt* neutrons, the emission of some suffers a time delay of the order of several seconds. We shall see in §23 that this delayed neutron emission in fission products is of absolutely fundamental importance in the control of nuclear reactors. As an example of a fission product that emits delayed neutrons in coincidence with a β decay we may quote ^{87}Br, which decays with a half-life of 55 s. Likewise many delayed proton emitters are formed in nuclear reactions, e.g. ^{25}Si, with a half-life of 0.22 s.

The fact that the very much slower β decay can precede the nucleon emission can only mean that the prior emission, but not the subsequent emission, must be energetically impossible. To investigate the necessary conditions for this situation to prevail we first note that the prior emission of a nucleon will be impossible if the first nucleus is formed in its ground state and lies between the drip lines. (We are ignoring in this argument the occasional nucleus lying between the drip lines which, because of fluctuations with respect to the smooth semi-empirical formula, will be a nucleon emitter even in its ground state.) But then the daughter nucleus formed by β decay will also lie between the drip lines, so that it too will be unable to emit a nucleon from its ground state. Thus delayed nucleon emission is possible only if the β decay can feed a sufficiently high-lying state of the second nucleus†.

The emitted nucleon will have maximum kinetic energy when the excited state of the daughter nucleus which the β decay feeds is so high that the electron–neutrino pair carries away zero kinetic energy. Thus for delayed neutron emission we have

$$T_{\mathrm{n}} \leqslant (M_{\mathrm{nuc}}(Z,N) - M_{\mathrm{nuc}}(Z+1,N-2) - M_{\mathrm{n}} - m_{\mathrm{e}})c^2 \qquad (17.16a)$$

while for delayed proton emission

$$T_{\mathrm{p}} \leqslant (M_{\mathrm{nuc}}(Z,N) - M_{\mathrm{nuc}}(Z-2,N+1) - M_{\mathrm{p}} - m_{\mathrm{e}})c^2. \qquad (17.16b)$$

†The phenomenon of delayed nucleon emission is somewhat paradoxical, in that the nucleus in question can emit a nucleon only if accompanied by an electron–neutrino pair. We can understand this by noting that the nucleon emission occurs only if there becomes available for it a part of the energy that would have been carried away entirely by the electron–neutrino pair, had the β decay fed the ground state of the daughter nucleus.

In terms of the mass excesses this becomes

$$T_{\mathrm{n}} \leqslant \Delta(Z, N) - \Delta(Z+1, N-2) - \Delta_{\mathrm{n}} \qquad (17.17\mathrm{a})$$

and

$$T_{\mathrm{p}} \leqslant \Delta(Z, N) - \Delta(Z-2, N+1) - \Delta_{\mathrm{H}} - 2m_{\mathrm{e}}c^2. \qquad (17.17\mathrm{b})$$

The condition for the delayed, but not the direct, emission of neutrons to be possible is now seen to be

$$M_{\mathrm{nuc}}(Z, N-1) + M_{\mathrm{n}} > M_{\mathrm{nuc}}(Z, N) > M_{\mathrm{nuc}}(Z+1, N-2) + M_{\mathrm{n}} + m_{\mathrm{e}} \qquad (17.18\mathrm{a})$$

i.e.

$$\Delta(Z, N-1) + \Delta_{\mathrm{n}} > \Delta\,(Z, N) > \Delta(Z+1, N-2) + \Delta_{\mathrm{n}}. \qquad (17.19\mathrm{a})$$

In terms of the internal energy per nucleon e this condition becomes

$$(A-1)e(Z, N-1) > Ae(Z, N) > (A-1)e(Z+1, N-2) - T_{\mathrm{on}} \qquad (17.20\mathrm{a})$$

where $T_{\mathrm{on}} = 782$ keV is defined in equation (16.11). Likewise for the delayed, but not the direct, emission of protons to be possible, the condition is

$$M_{\mathrm{nuc}}(Z-1, N) + M_{\mathrm{p}} > M_{\mathrm{nuc}}(Z, N) > M_{\mathrm{nuc}}(Z-2, N+1) + M_{\mathrm{p}} + m_{\mathrm{e}} \qquad (17.18\mathrm{b})$$

i.e.

$$\Delta(Z-1, N) + \Delta_{\mathrm{H}} > \Delta(Z, N) > \Delta(Z-2, N+1) + \Delta_{\mathrm{H}} + 2m_{\mathrm{e}}c^2 \qquad (17.19\mathrm{b})$$

or equivalently

$$(A-1)e(Z-1, N) > Ae(Z, N) > (A-1)e(Z-2, N+1) + T_{\mathrm{on}} + 2m_{\mathrm{e}}c^2. \qquad (17.20\mathrm{b})$$

We now wish to express these conditions in terms of the simple mass formula (14.12), which we use in the form (14.29), proceeding exactly as we did above for the case of direct nucleon emission. The conditions for no direct emission reduce, of course, to the inequalities (17.9a) and (17.9b) *not* being satisfied. Assuming this to be the case, i.e. that our nucleus lies within the drip lines, equation (17.20a) for delayed neutron emission becomes

$$Ae_0(A) - (A-1)e_0(A-1) + 2a_{\mathrm{sym}}\left(\frac{(Z - Z_0(A))^2}{Z_0(A)} - \frac{(Z+1-Z_0(A-1))^2}{Z_0(A-1)}\right) + T_{\mathrm{on}} > 0 \qquad (17.21\mathrm{a})$$

while equation (17.20b) for delayed proton emission becomes

$$Ae_0(A) - (A-1)e_0(A-1) + 2a_{\mathrm{sym}}\left(\frac{(Z - Z_0(A))^2}{Z_0(A)} - \frac{(Z-2-Z_0(A-1))^2}{Z_0(A-1)}\right) - T_{\mathrm{on}} - 2m_{\mathrm{e}}c^2 > 0. \qquad (17.21\mathrm{b})$$

To see the significance of the conditions (17.21) we proceed in exactly the same way as we treated the conditions (17.9) for direct nucleon emission, making the same approximations and neglecting T_0 and $2m_ec^2$. We then find

$$e_0 - \frac{4a_{\text{sym}}}{Z_0}(Z - Z_0)\left(\pm\frac{3}{2} - \varepsilon\right) > 0 \qquad (17.22)$$

where the upper sign refers to delayed neutron emission and the lower to delayed proton emission. Thus there will be delayed neutron emission for

$$\begin{aligned} Z &< Z_0\left(1 + \frac{e_0}{2a_{\text{sym}}}\frac{1}{3 - 2\varepsilon}\right) \\ &\simeq Z_0\left(1 - \frac{0.2}{3 - 2\varepsilon}\right) \end{aligned} \qquad (17.23a)$$

and delayed proton emission for

$$\begin{aligned} Z &> Z_0\left(1 - \frac{e_0}{2a_{\text{sym}}}\frac{1}{3 + 2\varepsilon}\right) \\ &\simeq Z_0\left(1 + \frac{0.2}{3 + 2\varepsilon}\right). \end{aligned} \qquad (17.23b)$$

Both these conditions for *delayed* nucleon emission can be satisfied between the drip lines, i.e. without satisfying the conditions (17.14) for *direct* nucleon emission. It follows that there exists a zone of intermediate neutron richness in which we have stability with respect to direct neutron emission but not with respect to delayed neutron emission; similar remarks apply to proton-rich nuclei. Referring to figure 17.1, we see that each of these zones lies between the appropriate drip line (broken) and the adjacent dotted line†, which we call a *delayed drip line*. Between the dotted lines we shall have stability with respect to both direct and delayed nucleon emission, at least insofar as the Bethe–von Weizsäcker mass formula (14.12) is valid.

This last proviso implies that our boundaries of the delayed drip zones may not be very accurate, and indeed, the examples of delayed nucleon emitters that we have quoted, ^{87}Br and ^{25}Si, are found to lie just outside the appropriate zones of figure 17.1. The point to be made here is that delayed nucleon emission can be expected to be a quite general feature of the nuclear landscape, beyond a certain distance from the stability line.

†Because of the approximations made in going from (17.21) to (17.23) we have shown the dotted lines as corresponding to the limits defined by the former rather than the latter.

α emission. We consider now the emission of nuclei containing more than one nucleon. Returning to the general relation (17.3) we find that the condition for a nucleus of mass number A to emit a nucleus with n nucleons is

$$\Delta e \geqslant -\frac{n}{A}(e - e_n) \tag{17.24}$$

where e_n is the internal energy per nucleon in the emitted nucleus, and we have written $e(A-n) \equiv e$, $e(A) \equiv e + \Delta e$; we notice that (17.7) is a special case of this.

We look first at the possible emission of deuterons, i.e. ^{2}H nuclei, abbreviated as d. Since the internal energy per nucleon is - 1.11 MeV, the condition (17.24) becomes

$$\Delta e \geqslant -\frac{2}{A}(e + 1.11)\ \text{MeV}. \tag{17.25}$$

Now there is no absolute reason why this condition should not be satisfied, especially if we allow our nucleus to be in a high state of excitation. However, we see from (17.7) that this value of Δe is close to that required for the emission of both a proton and a neutron, either simultaneously or successively, the value of e being of the order of -6 to -8 MeV: because of the weak binding of the deuteron it is about as easy to emit an unbound deuteron. We conclude, then, that there would be far more energy available for single-nucleon emission, which would thus completely dominate any deuteron emission in most conceivable cases. Similarly negative remarks apply to the possible emission of tritons (^{3}H nuclei, abbreviated as t) and helions (^{3}He nuclei, abbreviated as τ), both being quite weakly bound ($e = -2.83$ MeV and -2.57 MeV, respectively).

The situation changes markedly when we pass to ^{4}He nuclei, these being much more strongly bound than the other light nuclei we have considered. In fact, their internal energy per nucleon of -7.07 MeV compares with that of most heavier nuclei, so that (17.24) will be much easier to satisfy, there being a strong cancellation on the right-hand side. This is the reason why α decay is such a widespread phenomenon, while lighter nuclei, other than single nucleons, are not emitted at all.

Writing (17.24) explicitly for α decay thus

$$\Delta e \geqslant -\frac{4}{A}(e + 7.07)\ \text{MeV} \tag{17.26}$$

we examine the implications for the ground states of nuclei lying on the stability line. Referring to figure 14.2, we see that Δe is negative for $A < 56$, while $e < -7.07$ MeV (provided $A > 4$, so that the nucleus is at least big enough to emit an α particle). Thus the right-hand side of (17.26) will always be positive, so that (17.26) cannot possibly be satisfied: there can be no α decay from the ground states of nuclei lying on the stability line for $A < 56$. On the other hand, Δe turns positive beyond $A = 56$, while the right-hand side of (17.26) becomes smaller and smaller, so that as we move towards heavier nuclei there will be an increasing likelihood of satisfying the condition (17.26). To see what happens quantitatively beyond $A = 56$ we use again the semi-empirical relation (17.8), taken from figure 14.2. We find that the condition (17.26) is satisfied for $A > 151$ (see Problem III.15). In fact, the lightest α emitter on the stability line is ^{148}Gd ($Q = 3.18$ MeV, $t_{1/2} = 93$ yr), but it is only beyond lead that α decay becomes a frequent phenomenon.

Spontaneous fission. We now consider the question of whether the emission of nuclei heavier than α particles from the ground states of nuclei lying on the stability line could ever be energetically possible. In the first place, we note that for nuclei with $A < 56$, just as α decay is impossible, so is any other conceivable fragmentation: the binding energy per nucleon in *both* the nuclei resulting from such a partition would always be less than in the original nucleus.

In the same way, we see that for $A > 112$, partitions in which *both* fragments have more than 56 nucleons will be energetically possible, since the binding energy per nucleon in both fragments will be *greater* than in the original. Actually, this argument overlooks the fact that the product nuclei will lie some distance from the stability line, so presumably one would have to go to somewhat higher values of A before such fragmentation became generally possible. In any case, because of barrier effects (discussed in the next section) it is only for much larger values of A that such *spontaneous fission*, as it is called, can proceed at an observable rate, and even then α decay will often dominate, despite there being less energy available. For example, in the case of ^{238}U the Q value for spontaneous fission is about 190 MeV in the symmetric mode, i.e. when the two fragments are identical, while the Q value for α decay is only 4.27 MeV; nevertheless, the latter process is some two million times more probable.

To examine in more detail the energetics of fission we return to the semi-empirical mass formula (14.12) and note that the only terms that can change are the surface and Coulomb terms. We find maximum energy release for the symmetric mode (Problem III.16a), but it should be remarked that the surface and Coulomb contributions have opposite signs, with the former tending to prevent fission: fission leads to an increase in the total nuclear surface area. Clearly, insofar as fission occurs, the Coulomb term† must dominate, and from equation (17.8) we see that the energy release in the

†Thus, as Feynman has pointed out, energy derived from fission can be said to be basically electrical. Indeed, the specifically *nuclear* contribution to the energy release, i.e. the surface term, is negative, representing a *loss* of energy.

symmetrical mode varies roughly as A^2 (see Problem III.17). On the other hand, equation (17.8) also shows that the energy available for α decay increases only linearly with A (see Problem III.18), so that spontaneous fission must become increasingly important compared with α decay as we move towards heavier nuclei, since barrier penetration increases very rapidly with the available energy (see the next section). It is, in fact, spontaneous fission, rather than α decay, that sets an upper limit on the mass numbers of nuclei that hold together long enough to be observable.

It should be realised that fission is not necessarily symmetrical; indeed the symmetrical mode is not the most common one, even though it is energetically the most favoured mode (within the framework of the mass formula of Bethe and von Weizsäcker). Nevertheless, fission modes in which one of the two final nuclei has fewer than 56 nucleons are very rare, unless, of course, one regards α decay as a special case of spontaneous fission (usually one does not, but this is a matter of semantics). For example, the fission mode of ^{234}U corresponding to the emission of ^{16}O is energetically possible, but because of barrier effects it is completely dominated by four successive α decays, as we shall discuss in the next section. However, the emission of ^{14}C by ^{223}Ra has been observed recently (1984); it is easy to check that more energy is released in this decay mode than in the emission of ^{12}C (see Problem III.20).

It remains to discuss the nuclei $56 < A < 112$. We have already noted that α decay of these nuclei will be impossible, but it is rather difficult to make any general statement about the energetics of the emission of heavier nuclei. However, we can resort to the mass tables and consider specific cases. It will then be found, for example, that the symmetric fission of ^{96}Mo is energetically possible, although only by 200 keV, which is far too small for the fission barrier to be penetrated at a measurable rate. It turns out, in fact, that for all practical purposes all nuclei in this mass-number range (actually, from 1 to 148) and lying on the β stability line can be regarded as completely stable with respect to fragmentation processes.

§18 FRAGMENTATION: BARRIER PENETRATION

If a nucleus is energetically capable of disintegrating the question arises as to why this does not take place the instant it is formed, rather than over a finite timescale. Considering the specific case of α decay, we must surmise that the first stage in this process will be the momentary clustering together of two protons and two neutrons in the surface of the nucleus to form something resembling an α particle. Let us denote the number of times per second such a formation appears as f. Already we see one reason why the decay process should take a finite time, following the radioactivity law (10.1). However, we would not expect f to vary very much from one nucleus to another, while in fact α lifetimes, like β lifetimes, vary over an enormous range (see table 18.1). This difficulty is related to another paradox, known from the early days of α decay and which we now discuss.

Table 18.1 Some typical α emitters†.

Nucleus	$t_{1/2}$	Q_α(MeV)	$\log_{10} \nu$
^{148}Gd	93 yr	3.27	50
^{211}Po	0.56 s	7.59	51
^{212}Po	3×10^{-7} s	8.96	53
^{221}Fr	4.8 min	6.46	54
^{224}Ra	3.6 d	5.79	55
^{232}Th	1.4×10^{10} yr	4.05	57
^{233}U	1.6×10^{5} yr	4.91	57
^{235}U	7.1×10^{8} yr	4.68	55
^{238}U	4.5×10^{9} yr	4.27	60
^{239}Pu	2.4×10^{4} yr	5.24	59
^{241}Pu	15 yr	4.90	63

†Taken from the compilation of Rytz A 1973 *Atomic Data and Nuclear Data Tables* **12** 479

When an embryonic α particle forms in the nuclear surface, there will be a net attraction between it and the rest of the nucleus, because of the specifically nuclear forces. However, once the α particle has escaped, the nuclear interactions between it and the residual nucleus will cease to act and we shall be left with a purely Coulomb *repulsion*

$$V(r) = \frac{Z_1 Z_2 e^2}{r} \qquad r > R \tag{18.1}$$

where we have represented the charge of the α particle by Z_1 (so that the theory will be applicable to the emission of any charged particle), Z_2 is the charge of the *residual* nucleus and r the distance between it and the emitted particle. We suppose, furthermore, that the nuclear interactions are negligible immediately beyond the nuclear surface, radius R. On the other hand, in depicting this situation in figure 18.1, we have supposed that the attractive nuclear forces are completely dominant, even just within the nuclear surface, and can be represented by a very deep potential well. In reality, the transition from attraction to repulsion in the vicinity of the nuclear surface will be less abrupt than we have drawn it, but such details are of no concern here. Moreover, we do not have to specify what happens to $V(r)$ well within the nucleus; indeed, it will not be possible to define it there, since the α particle will not have a well defined identity in this region.

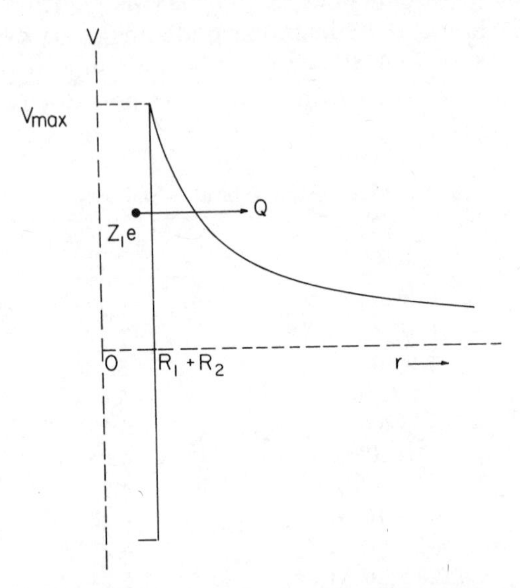

Figure 18.1 Coulomb barrier for charged particle emission.

The kinetic energy of relative motion of the emitted particle and the residual nucleus will, at any point r, be given in terms of the total decay energy Q_α by

$$K(r) = Q_\alpha - V(r) \tag{18.2}$$

since the Q value is just the relative kinetic energy as measured in the laboratory when the two nuclei are far apart, $V(r)$ then being zero. The paradox to which we have alluded now consists in the fact that α decay energies Q_α are always smaller than the maximum value of $V(r)$, as given by

$$V_{\max} = \frac{Z_1 Z_2 e^2}{R_1 + R_2} = \frac{1.44}{r_0} \frac{Z_1 Z_2}{A_1^{1/3} + A_2^{1/3}} \text{ MeV} \tag{18.3}$$

for any reasonable value of the nuclear radius parameter, r_0 (measured in fm). Thus, for ^{238}U we find $V_{\max} \simeq 30$ MeV, while $Q_\alpha = 4.27$ MeV. Classically, such a situation is impossible, since it means, according to (18.2) that the kinetic energy will be negative for some values of r; the α particle should remain for ever trapped inside the potential barrier that exists at the surface of the nucleus.

A classically forbidden barrier penetration of this sort is, of course, familiar to all students of elementary quantum mechanics. The relevance to the α decay paradox was first perceived in 1928 by Condon and Gurney and independently by Gamow. Their theory represented the first application of the newly developed quantum mechanics to nuclear physics. The point is that if an α particle of given energy Q arrives at the inside of the barrier of height given by (18.3) there will be a finite probability $T(Q)$, known as the *transmission coefficient*, that it will be able to 'leak' through the barrier and so escape. Thus the decay constant, i.e. the probability of decay per unit time, will be just

$$\lambda = fT(Q). \tag{18.4}$$

Now we cannot easily estimate f, the number of times per second that an α particle forms in the surface and tries to penetrate the barrier; all we can say is that we do not expect it to vary drastically from one nucleus to another. Thus the most one should attempt to do in any elementary discussion is to establish the approximate energy dependence of the transmission coefficient $T(Q)$, and see if we can account in this way for the very wide range of α lifetimes.

Very approximately $T(Q)$ varies according to the so-called *Gamow factor*†

$$T \propto \exp(-2\pi n) \tag{18.5}$$

where we have introduced the *Sommerfeld parameter*

$$n = \frac{Z_1 Z_2 e^2}{\hbar v} \tag{18.6}$$

v being the asymptotic relative velocity between the emitted particle and the residual nucleus

$$v = \left(\frac{2}{M}\frac{Q}{A}\right)^{1/2} \tag{18.7}$$

in which

$$A = \frac{A_1 A_2}{A_1 + A_2}. \tag{18.8}$$

Thus we can write

$$2\pi n = \gamma Z_1 Z_2 \sqrt{\frac{A}{Q}} \tag{18.9}$$

where

$$\gamma = \frac{\pi e^2}{\hbar}(2M)^{1/2} = 0.99\ (\mathrm{MeV})^{1/2}. \tag{18.10}$$

†For a derivation see, for example, p. 123 of *Quantum Mechanics* by E Merzbacher (1961) (New York: Wiley). Our expression for the transmission coefficient is valid for arbitrary charges of the two nuclei involved. This is particularly important in view of the fact that it applies not only to two nuclei that are separating but also to two that are approaching, and thus is relevant to nuclear reactions as well as decay processes (see Chapter IV).

The constant of proportionality in (18.5) will depend on the radius $R_1 + R_2$ of the barrier and on the details of $V(r)$ for $r < R_1 + R_2$; it will also depend on the energy Q, but much less strongly than does the Gamow factor, so that we can combine (18.4) and (18.5) to obtain

$$\lambda = \nu e^{-2\pi n} \tag{18.11}$$

where to a very rough approximation ν is the same for all decaying nuclei. Then for the specific case of α decay we have

$$\ln \lambda = \ln \nu - 4\gamma \frac{Z-2}{Q_\alpha^{1/2}} \tag{18.12}$$

since $Z_1 = 2$, $A \simeq 4$ (all α emitters being heavy), and $Z_2 = Z - 2$, Z being the atomic number of the emitter.

A measure of the validity of this highly simplified formula† may be had from table 18.1, where we have extracted from equation (18.12) the value of ν for a number of typical α emitters. The values of this 'constant' vary by a factor of 10^{12}, but it must be realised that the lifetimes shown vary by a factor of 10^{29}. Thus the Gamow factor accounts at least for a large part of this enormous range of lifetimes; the remaining part, as represented by the variation of ν, contains important nuclear structure information.

It must be emphasised how enormously sensitive the α decay lifetimes are to the available energy, because of the Gamow factor. With Z in the region of 90, a typical value for α emitters, a change in Q_α from 4 to 9 MeV changes the lifetime by a factor of 10^{25}, which corresponds to a lifetime of 10^{-7} s becoming 10^{11} yr. Again, if the α decay energy of ^{235}U were increased by a mere 50 keV, from 4.68 to 4.73 MeV, the half-life would be reduced by a factor of about 2.4, i.e. to about 3.0×10^8 yr. Since all the nuclei in the earth were synthesised about 4.5×10^9 years ago it follows that the amount of ^{235}U in the earth would be only 2×10^{-3} of what it actually is at the present time, and its concentration in natural uranium correspondingly weaker, assuming that the lifetime of the dominant ^{238}U had not also been changed (see Problem III.21) The implications of this are enormous, since it means that a fission chain reaction would have been very much more difficult, maybe even impossible, to establish at the present time, even though it would still have been possible at a sufficiently remote time in the past (see Chapter V).

The Gamow factor is also extremely sensitive both to the charge and the mass of the emitted particle, and tends to discriminate very strongly against the emission of heavy particles. For example, the Q value for emission of ^{16}O nuclei by ^{234}U is 34.51 MeV, while only 4.86 MeV is available for α decay. Nevertheless, we find the former process to be less probable by a

†See Problem III.21 for another formula.

factor of 10^{127} (assuming ν to be the same for both processes†), which means that it will be totally negligible, the alpha half-life being 2.5×10^5 yr. In fact what happens is that ^{234}U emits a cascade of four α particles in succession, rather than emitting a single ^{16}O nucleus (see Problem III.19).

Spontaneous fission. When a nucleus splits into two portions of comparable size, it is clear that we are dealing with a mechanism that is radically different from that of α decay: the latter involves merely a momentary clustering in the surface, while fission requires that the entire nucleus participates collectively. Pictorially, the different stages leading up to fission could be visualised as in figure 18.2. The originally undeformed nucleus (i) first undergoes a spontaneous elongation (ii), a neck develops (iii) and then ruptures, so that we are left with two nuclei moving apart under the influence of their mutual Coulomb repulsion (iv) and (v).

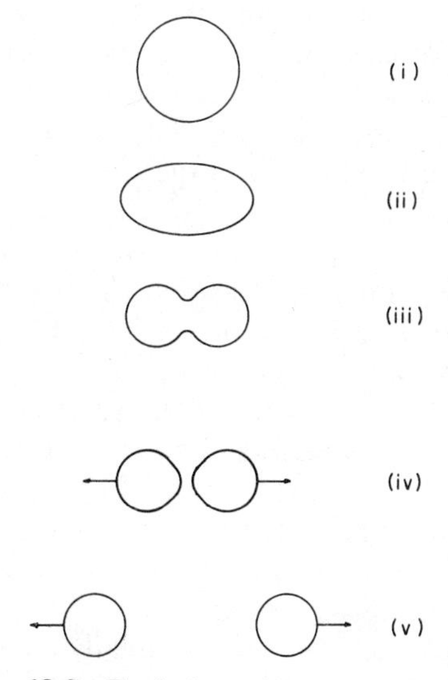

Figure 18.2 Evolution of fissioning nucleus.

Let us suppose that already at the instant of separation all nuclear interaction between the two nuclei has ceased, so that the net potential energy of interaction at this point will be purely Coulombic:

$$V_C = \frac{Z_1 Z_2 e^2}{R_1 + R_2}. \tag{18.13}$$

†Actually, ν will be much lower for ^{16}O emission, since the chance formation of ^{16}O clusters in the nuclear surface will be much rarer than for ^4He clusters.

For symmetric fission of ^{238}U this gives about 220 MeV, while the kinetic energy of relative motion when the two product nuclei are far apart is only 190 MeV, this being the Q value for this fission process. Thus once again it is apparent that we have a barrier penetration not totally dissimilar from that shown in figure 18.1 for the emission of light charged particles. More generally, we might conclude that a barrier penetration is involved whenever spontaneous fission takes a measurable time to occur, rather than taking place instantaneously upon formation of the nucleus.

For $r < R_1 + R_2$ it is not possible to define a potential of interaction $V(r)$ between the two product nuclei because in that region they do not yet have a well defined identity. Nevertheless, if we assume that the changes in configuration of the fissioning nucleus sketched in figure 18.2 take place very slowly, then the total energy of each instantaneous configuration defines a potential energy function. This is the quantity that we plot in figure 18.3, the variable r with values lower than $R_1 + R_2$ just being interpreted as a parameter characterising the state of evolution of the fissioning nucleus. Beyond $R_1 + R_2$ this potential function will become simply the usual Coulomb repulsion, provided the zero is taken as the total internal energy of the two product nuclei after they have separated. With this convention we see also that the value of V for zero deformation, corresponding to the initial stage (i) in figure 18.2, i.e. $V(r = 0)$, will be just equal to the Q value of the fission process.

Now the semi-empirical mass formula can be used to calculate the binding energies of the initial nucleus and the two final nuclei, and thus serves to determine $V(r = 0) = Q$. However, it cannot be used as it stands to calculate the changes in binding energy as the initial nucleus goes through the deformation sequence of figure 18.2. Nevertheless, we can take the semi-empirical formula as a guide and assert that the energy *changes* can have only two sources: the surface term and the Coulomb term. Deformation leads inevitably to an increase in the surface area and hence in the surface energy, while the Coulomb energy must decrease, since the mean distance between the nucleons increases.

The Coulomb term must dominate eventually if the Q value is positive, i.e. if spontaneous fission is energetically possible, but if changes in the surface term dominate for small deformations, then $V(r)$ will at first rise and we shall have the fission barrier shown in figure 18.3. Actually, at the point $r = R_1 + R_2$ the surface term will still not have reached its limiting value since the two product nuclei will still be deformed under the influence of their mutual attraction. Thus V_{max} will be considerably smaller than the purely Coulomb value (18.13), and likewise for the barrier height

$$E_b = V_{max} - Q. \qquad (18.14)$$

In the case of the uranium isotopes, for example, the barrier height is of the order of 6 MeV, rather than the 30 MeV or so that one would estimate on the basis of (18.13). Indeed, it is conceivable that there is no barrier at all. This will happen if the surface term does not dominate even for small deformations, since then $V(r)$ will decrease monotonically, as shown by the broken curve in figure 18.3. Such nuclei will undergo spontaneous fission the instant they are formed, and will never be observed.

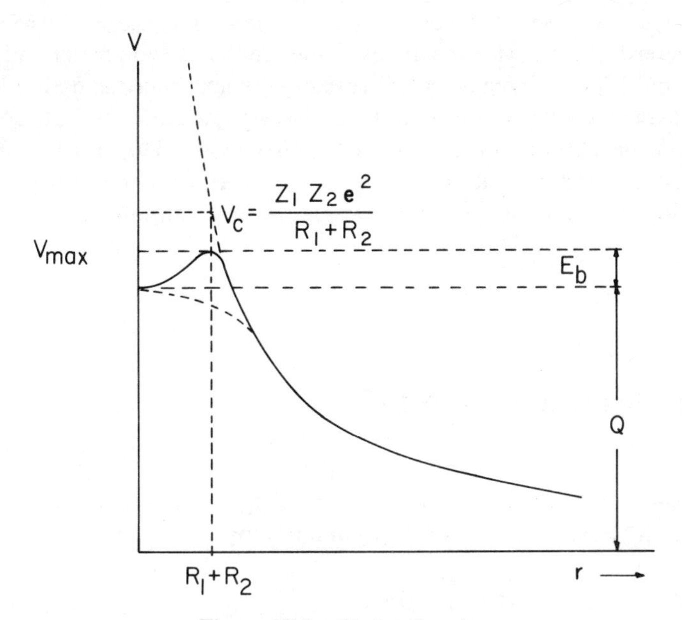

Figure 18.3 Fission barrier.

To investigate further this possibility of there being no fission barrier, let us denote by E_C and E_{surf} the total Coulomb and surface energies, respectively, of the original undeformed nucleus

$$E_C = a_C \frac{Z^2}{A^{1/3}} \tag{18.15}$$

$$E_{surf} = a_{surf} A^{2/3} \tag{18.16}$$

where a_C and a_{surf} are just the usual parameters of the semi-empirical formula. It may be shown† that if $E_C > 2E_{surf}$ then for a small deformation the *change* in the Coulomb term will be greater than the *change* in the surface term, and there will be no fission barrier. It is convenient therefore to

†See, for example, p. 570 of the book by Preston and Bhaduri quoted in the General Bibliography.

introduce the *fissility parameter*

$$x = \frac{E_C}{2E_{surf}} = \frac{a_C}{2a_{surf}} \frac{Z^2}{A}$$

$$\simeq \frac{1}{46} \frac{Z^2}{A} \qquad (18.17)$$

where we have used our values $a_C = 0.6$ MeV and $a_{surf} = 13.9$ MeV. This parameter increases with A, and as it does so the fission barrier will become lower and lower, giving a second reason why spontaneous fission lifetimes should decrease as we move towards heavier nuclei†. In fact, for nuclei lying on the stability line x will reach unity by $A = 330$, so this constitutes an absolute upper limit on the mass numbers of nuclei that can exist, assuming that our model is not too grossly oversimplified.

PROBLEMS FOR CHAPTER III

III.1 Show that the recoil energy of the daughter nucleus of a high-energy β decay ($Q \gg m_e c^2$) is given approximately by

$$T_{recoil} \simeq \frac{Q^2}{2AMc^2}$$

at the end point of the spectrum.

For the case of the β^- decay of ^{12}B, calculate the shift in the end-point energy of the electron spectrum associated with the nuclear recoil.

III.2 Show that ^{151}Sm β decays into ^{151}Eu, even though it is more strongly bound.

III.3 Show that ^{64}Cu can undergo both β^- and β^+ decays.

III.4 Which isobars $A = 92$ are β stable? For the three isobars $A = 92$ closest to the stable ones, indicate the form of instability, i.e. $\beta^{\pm}$ or electron capture, and calculate the maximum energy of the emitted electron or neutrino.

†The first reason being that the Q values increase with A, as noted in the previous section. Even for a fixed barrier height we would expect the penetration to increase with the energy Q with which the product nuclei emerge beyond the barrier (note that the Gamow factor will no longer be applicable, V_{max} no longer being given by (18.13)).

III.5 Show that there are three stable isobars with $A = 130$.

III.6 Show that neutral atoms of ^{187}Re can undergo β^- decay, but that the bare nuclei are stable.

Is it possible that there are bare nuclei that undergo β^+ decay while the corresponding neutral atoms are stable?

III.7 Hydrogen gas is compressed until 95% of the protons have captured electrons and been transformed into neutrons. What will its density then be? Take the possibility of *nucleon* degeneracy into account.

III.8 What is the minimum energy of neutrinos needed to initiate the reactions

(*a*) $\quad {}^{37}\mathrm{Cl} + \nu \rightarrow {}^{37}\mathrm{A} + \mathrm{e}^-$

(*b*) $\quad {}^{71}\mathrm{Ga} + \nu \rightarrow {}^{71}\mathrm{Ge} + \mathrm{e}^-$?

III.9 Derive equation (17.10), making reasonable approximations.

III.10 Starting from the mass formula (14.12), and defining Δe as in equation (17.7), show that for neutron emission

$$(\Delta e)_\mathrm{n} = -\frac{4}{3}\frac{a_\mathrm{C} Z^2}{A^{7/3}} + \frac{4a_\mathrm{sym}}{A}\frac{Z}{A}\left(\frac{N-Z}{A}\right) - \frac{1}{3}\frac{a_\mathrm{surf}}{A^{4/3}}$$

while for proton emission

$$(\Delta e)_\mathrm{p} = \frac{a_\mathrm{C} Z^2}{A^{7/3}}\left(\frac{2A}{Z} - \frac{4}{3}\right) + \frac{4a_\mathrm{sym}}{A}\frac{N}{A}\left(\frac{Z-N}{A}\right) - \frac{1}{3}\frac{a_\mathrm{surf}}{A^{4/3}}.$$

III.11 Using the results of the previous problem show that when the initial nucleus is on the stability line

$$\Delta e = \frac{2}{3} a_\mathrm{C} \frac{(Z_0(A))^2}{A^{7/3}} - \frac{1}{3}\frac{a_\mathrm{surf}}{A^{4/3}}$$

for both neutrons and protons. (Note that it is only along the stability line that Δe is the same for both neutrons and protons.)

III.12 Show that the separation energy for nucleons is given by

$$S_\mathrm{N} = -(e + A\,\Delta e).$$

III.13 Using the results of the two previous problems show that the nucleon separation energy along the stability line is given by

$$S_\mathrm{N} = a_\mathrm{vol} - \frac{2}{3}\frac{a_\mathrm{surf}}{A^{1/3}} - a_\mathrm{C}\frac{Z_0}{2A^{1/3}}\left(1 + \frac{4}{3}\frac{Z_0}{A}\right).$$

Calculate the separation energy of a nucleon from ^{236}U and compare with experiment for both neutron and proton cases (the neglect of the pairing term in the mass formula (14.12) accounts in part for the disagreement).

III.14 Using the mass tables, show that neither ^{87}Br nor ^{25}Si can emit nucleons directly, but that they can emit β delayed nucleons. In each case calculate the maximum energy of the emitted nucleons.

III.15 Assuming equation (17.8) to be valid, determine the smallest A value for which a nucleus on the line of β stability can emit an α particle from the ground state.

III.16 (*a*) Show from the mass formula (14.12) that the energy release in spontaneous fission is maximal for the symmetric mode.

(*b*) Using this formula calculate the energy release in the symmetric fission of ^{238}U.

(*c*) Using the mass tables, show that ^{96}Mo is energetically capable of undergoing spontaneous fission.

III.17 Using the approximation of equation (17.8), show that the energy release for symmetric fission increases as A^2.

III.18 Using the approximation of equation (17.8), show that the energy release in α decay increases linearly with A.

III.19 Show from the mass tables that ^{234}U is energetically capable of emitting ^{16}O nuclei.

III.20 Show that more energy is released in the emission of ^{14}C from ^{223}Ra than in the emission of ^{12}C. Try to explain this result physically.

III.21 Given the data of table 18.1 on the α decay of ^{235}U, what would the half-life have been if Q_α had been equal to 4.73 MeV?

What would the percentage concentration of ^{235}U in natural uranium have been at the present time with this modified value of Q_α, if the properties of ^{238}U were not modified? (The actual concentration of ^{235}U at the present time in the solar system is 0.7%; this uranium was synthesised 4.5×10^9 years ago.)

III.22 The following empirical formula has been given for the α half-lives of even–even nuclei

$$\log_{10} t_{1/2}(\mathrm{s}) = \frac{2.11329Z - 48.9879}{\sqrt{Q_\alpha(\mathrm{MeV})}} - 0.39004Z - 16.9543$$

(Viola and Seaborg 1961 *J. Inorg. Nucl. Chem.* **28** 741).
Use this formula to calculate $t_{1/2}$ for all the nuclei listed in table 18.1 (including those that are not even–even), and in each case compare with experiment.

CHAPTER IV

NUCLEAR REACTIONS

§19 GENERAL INTRODUCTION TO NUCLEAR REACTIONS

In the preceding chapter we discussed the spontaneous transmutations that an isolated nucleus can undergo. When we have a collision between two nuclei, the range of possibilities is widened enormously: not only can we bring together any pair of nuclei that we wish but energy can be brought in to permit processes which would not otherwise be possible.

That the bombardment of one nucleus by another could lead to their transmutation was first shown by Rutherford in 1919, when, in the simplest experiment imaginable, he observed that the passage through air of α particles from a natural radioactive source led to the production of protons. The process responsible was subsequently identified as the reaction

$$\alpha + {}^{14}\mathrm{N} = {}^{17}\mathrm{O} + \mathrm{p}. \tag{19.1}$$

(The alternative, and somewhat simpler, process

$$\alpha + {}^{14}\mathrm{N} = {}^{13}\mathrm{C} + \alpha + \mathrm{p} \tag{19.2}$$

in which a proton is simply knocked out by an α particle, was eliminated as a possible explanation simply by counting tracks in cloud-chamber photographs.)

In the years following this discovery it was found that nearly all the light elements up to potassium could be made to emit protons under bombardment by α particles from natural sources. However, neutrons also could be emitted in such reactions as

$$\alpha + {}^{11}\mathrm{B} = {}^{14}\mathrm{N} + \mathrm{n}. \tag{19.3}$$

It was in this way, in fact, that Chadwick in 1932 first identified the neutron, whose existence had been suspected for some time.

Nevertheless, despite this great success, the use of radioactive samples as the source of projectiles constituted a severe limitation for several reasons: only α particle beams were possible, the available intensities were very low, and the projectile energies were likewise inconveniently low. This last point is particularly important in view of the Coulomb repulsion between the projectile and target nuclei, since there will be a barrier, similar to the one discussed in connection with α decay in §18, that will have to be penetrated or surmounted before any specifically nuclear interaction between the two nuclei can take place. Failing this, the two nuclei will simply undergo *elastic scattering*, i.e. the incident nucleus may change direction, and the target nucleus recoil, but with both remaining structurally unchanged. This problem of Coulomb inhibition will become worse with heavier target nuclei, because of their higher Z value, and requires higher bombarding energy if it is to be overcome (see §20).

The advent of accelerators marked, then, a most significant advance, since it permitted not only much greater beam intensities and energies, but also meant that one was no longer restricted to α particles as projectiles. The first completely artificial transmutation of a nucleus was achieved in 1932 by Cockcroft and Walton, who used their electrostatic accelerator to show that ^{7}Li disintegrated into two α particles when bombarded by protons of 125 keV or more, thus†

$$\mathrm{p} + {}^7\mathrm{Li} = {}^4\mathrm{He} + {}^4\mathrm{He}. \qquad (19.4)$$

(Protons encounter a lower Coulomb barrier than do α particles and so will initiate reactions at lower bombarding energies, generally speaking.)

With later developments in accelerator technology it has become possible for all nuclei to be transmuted in some way or other. In the first place beams of higher and higher energy and intensity are being obtained, and secondly, heavier and heavier nuclei are being accelerated. One obvious source of interest in these so-called *heavy-ion* reactions lies in the possibility they offer of synthesising nuclei far from the stability line.

However, reactions in which the projectile is something other than a nucleus are also of great importance. One such case is that of reactions initiated by neutrons. Of course, in studying these in the laboratory, it must be realised that neutrons cannot themselves be accelerated, but they can be obtained as a product of some charged-particle reaction, which itself will require an accelerator beam; a convenient source of neutrons consists of the reaction

$$\mathrm{d} + \mathrm{t} = \alpha + \mathrm{n}. \qquad (19.5)$$

Also of considerable interest are the so-called *photonuclear* reactions, in which the projectile consists of photons. In these reactions the absorbed

† It is of no consequence whether we write α or ^{4}He in this equation.

photon communicates all its energy to the target nucleus, which may be induced to emit nucleons, or heavier particles, e.g.

$$\gamma + {}^{16}\mathrm{O} \to \begin{cases} {}^{15}\mathrm{N} + \mathrm{p} \\ {}^{12}\mathrm{C} + \alpha. \end{cases} \tag{19.6}$$

These reactions are important as a tool for studying nuclear structure, and also play a significant role in stellar nucleosynthesis (Chapter VI).

As for the laboratory production of a photon source we note that photons, like neutrons, cannot be obtained directly as an accelerator beam, but they too can be produced in a charged-particle reaction, e.g. a reaction such as

$$\mathrm{p} + \mathrm{t} = \alpha + \gamma. \tag{19.7}$$

The secondary radiations of electron accelerators also constitute useful photon sources.

Reactions in which the projectile consists of electrons have also been extensively studied; electron beams can, of course, be produced directly by an accelerator. These reactions are similar to photonuclear reactions in that they consist essentially of the communication of energy via the electromagnetic interaction to the target nucleus, the electron being conserved, of course†. As a tool for the study of nuclear structure these reactions complement photonuclear reactions.

Finally, we should mention reactions induced by charged-pion beams. While of fundamental importance to both nuclear and elementary particle physics, they will not be considered here.

Having discussed the various projectiles that initiate reactions we now turn to the products. Most of the reactions that we have mentioned so far lead to just two product nuclei, but it is possible for more to be formed, as in reaction (19.2), and we denote the most general possible reaction by

$$\mathrm{A} + \mathrm{a} = \mathrm{B} + \mathrm{b}_1 + \mathrm{b}_2 + \ldots + \mathrm{b}_n + Q \tag{19.8}$$

or, in a more compact notation, by $\mathrm{A}(\mathrm{a}, \mathrm{b}_1 \ldots \mathrm{b}_n)\mathrm{B}$. Here A represents the target nucleus and a the projectile, while $\mathrm{B}, \mathrm{b}_1, \ldots, \mathrm{b}_n$ represent the product nuclei. Actually, the distinction between the nucleus B and the nuclei b_i is somewhat arbitrary, B simply denoting the heaviest nucleus formed and the b_i the various nuclear fragments; indeed, the distinction often cannot be made. The quantity Q in equation (19.8) represents the energy released in the reaction; see below for more details. Often the Q value is not displayed explicitly, as in all the foregoing examples.

Actually, for low bombarding energies it is rare to form more than two product nuclei, and even processes in which more than two are formed can

† We do not regard the weak interaction process of electron or neutrino capture (§11) as a reaction.

frequently be regarded as a simple two-nucleus reaction

$$A + a = B' + b_1 \tag{19.9a}$$

followed by a radioactive decay

$$B' = B + b_2 \tag{19.9b}$$

for example. Thus we shall seldom have to deal with anything more complicated than the two-body reaction

$$A + a = B + b + Q. \tag{19.10}$$

A most important exception to this rule is the so-called *neutron-induced fission*, which is distinguished from the spontaneous fission discussed in the last chapter in that the break-up of the heavy nucleus into two comparable nuclei occurs under the influence of neutron bombardment (it can also occur under the influence of charged-particle bombardment provided the energy is high enough, but with neutrons the fission may take place for very low bombarding energies). This process will be discussed in more detail in the next chapter, but the main point here is that it is a reaction of a form more complicated than that of equation (19.10), since a number of neutrons ('prompt' neutrons) may be emitted simultaneously with the break up into two large nuclei.

We have spoken of all the product particles B, $b_1, b_2, \ldots, b_n$ appearing in the general reaction (19.8) as being nuclei, but it is quite possible that one, or more, will be a photon, just as is the projectile a in the photonuclear reaction (19.6). Such photon emission occurs essentially through one of the product nuclei being formed in an excited state. However, because of the relative slowness of the subsequent γ decay it can be considered independently of the reaction proper, and it is not essential to regard the photon as being one of the reaction products.

A particularly important case of photon emission in nuclear reactions is that where only one product nucleus is formed

$$A + a = B + \gamma. \tag{19.11}$$

Such a reaction, of which (19.7) is an example, is known as *radiative capture*; it is the inverse of the photonuclear reaction.

It is not only through photon emission that surplus energy can be removed: another possibility is the creation of pions (see §8), but because of their finite mass ($\sim 270 m_e$) this can only occur for bombarding energies of several hundred MeV. Unlike photon production, however, this process cannot be considered independently of the reaction, since pions interact with nucleons through the strong nuclear force; we recall, in fact, that the interaction between nucleons is actually mediated by the exchange of pions. Here we shall not consider reactions for such high bombarding energies.

One possible outcome of the collision of two nuclei that we have already mentioned is elastic scattering, i.e. the same two nuclei emerge, both in their original states (usually the ground states):

$$A + a = A + a \tag{19.12}$$

but with a change of direction, so that in a laboratory experiment projectiles will be removed from the beam.

A related process is *inelastic scattering*, where one or other nucleus is left in an excited state, so that there will be a net loss in the translational kinetic energy of the two nuclei:

$$A + a = A^* + a. \tag{19.13}$$

This will be followed in general by a γ transition back to the ground state of A. (In some cases the projectile a can be excited also.)

Finally, to conclude this general discussion of reaction products, we note that for a given pair (A,a) of reacting nuclei there is often more than one set of possible reaction products

$$a + A \rightarrow \begin{cases} A + a \\ B + b_1 + \ldots \\ C + c_1 + \ldots \\ \text{etc.} \end{cases} \tag{19.14}$$

One example of this has already been seen in equation (19.6); another is

$$^{12}C + {}^{12}C \rightarrow \begin{cases} ^{12}C + {}^{12}C \\ ^{24}Mg + \gamma + 13.930 \text{ MeV} \\ ^{23}Na + p + 2.238 \\ ^{20}Ne + {}^{4}He + 4.616 \\ ^{23}Mg + n - 2.605 \\ ^{16}O + {}^{4}He + {}^{4}He - 0.114. \end{cases} \tag{19.15}$$

The first mode, elastic scattering, is always possible, and occasionally it is the only process possible, for instance when the bombarding energy is very low, or the colliding particles A and a are effectively without internal structure, as is the situation when the two nuclei are single nucleons and the bombarding energy is below the threshold for pion creation ($\simeq 300$ MeV).

It might appear from the foregoing discussion that nuclear reactions are primarily a laboratory phenomenon. However, there are some very important manifestations of nuclear reactions outside the nuclear physics laboratory. Terrestrially, for example, we have the neutron reactions, especially induced fission and radiative capture, occurring in the interior of nuclear reactors (both man-made and natural) and in nuclear weapons (see

Chapter V). Also to be noted are the reactions taking place in the upper atmosphere through cosmic ray bombardment, a most important case of which serves as the basis of the so-called 'radiocarbon' method of dating. Neutrons formed by primary cosmic ray bombardment react with atmospheric ^{14}N according to $^{14}N\,(n,p)\,^{14}C$. Now the ^{14}C that is formed is radioactive, undergoing β decay with a half-life of 5200 years, so that the ratio of ^{14}C to the stable ^{12}C in the atmosphere will reach a steady value when the rate of formation, which depends on the intensity of cosmic ray bombardment, is equal to the decay rate. Since living organisms take up CO_2 from the atmosphere they will have the same ratio of $^{14}C/^{12}C$, but after death no further exchange with the atmosphere is possible, and the concentration of the unstable ^{14}C will begin to decrease. Thus measurements of the $^{14}C/^{12}C$ ratio in a dead sample enable the date of death to be determined.

But it is on the astrophysical scale that the immense phenomenological significance of nuclear reactions becomes apparent, the energy output of all stable stars being derived from this source. By a stable star we mean one that is not contracting gravitationally (see Appendix D), and since the sun falls into this category the significance of nuclear reactions as the ultimate source of our own energy supplies will be appreciated. Furthermore, in generating energy through nuclear reactions the nuclei of the star will gradually be transmuted; it is in this way, in fact, that nearly all nuclei in the universe have been synthesised out of primordial hydrogen†. In a very real sense, then, nuclear reactions are the source of all the energy and all the matter (other than hydrogen) that man disposes of. (These questions are discussed more fully in Chapter VI.)

Energetics. The Q value of the reaction A(a, b)B is given in terms of the appropriate atomic masses by equation (A.5) as

$$Q = (M_A + M_a - M_B - M_b)c^2 \tag{19.16}$$

(for the case of radiative capture note that the photon mass is zero). Since the number of neutrons and the number of protons are both conserved in the reaction we may rewrite (19.16) in terms of the mass excess, thus

$$Q = \Delta_A + \Delta_a - \Delta_B - \Delta_b. \tag{19.17}$$

If Q is positive the reaction is said to be *exothermic* and can take place, in principle, even for vanishingly small bombarding energy, although the rate may become immeasurably small, because of the repulsive effect of the Coulomb barrier (see §20). If on the other hand Q is negative the reaction is said to be *endothermic* and cannot take place at all unless the bombarding

† A large fraction of the lightest nuclei were formed by reactions taking place in the original 'big bang'.

energy is in excess of a certain threshold value. In the frame of reference where the centre of mass of the entire system is at rest, the so-called CM (*centre of mass*) system (see Appendix A), this threshold energy is given simply as $E_{\text{CM}} = |Q|$. In the case of a laboratory experiment, where the target nucleus A is initially at rest, the threshold bombarding energy measured in the laboratory frame of reference becomes, according to equation (A.28)

$$E^0_{\text{lab}} = \frac{A_\text{a} + A_\text{A}}{A_\text{A}} |Q|. \tag{19.18}$$

The excess energy over and above Q is simply wasted in CM motion of the entire system.

Cross section. Consider now a single target nucleus A being bombarded by a beam of projectiles a, as shown in figure 19.1. The probability $\mathrm{d}P$ that a reaction will take place in time $\mathrm{d}t$ will be proportional to the flux density j of particles in the beam, i.e. to the number of particles that cross unit area perpendicularly in unit time. We write then

$$\mathrm{d}P = \sigma j \, \mathrm{d}t \tag{19.19}$$

where the constant of proportionality σ has the dimensions of area and is known as the *cross section*.

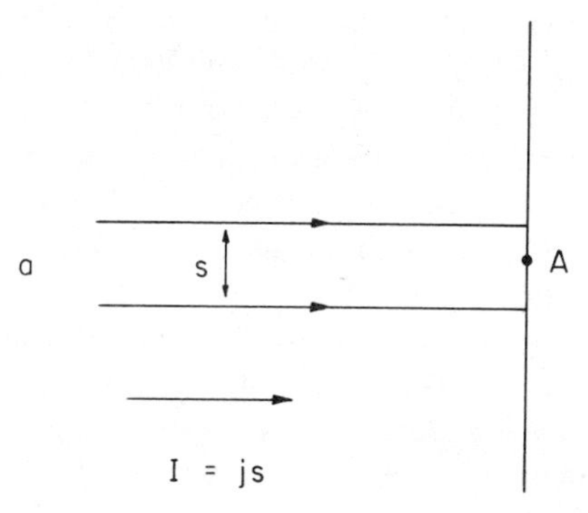

Figure 19.1 Beam of projectiles a bombarding target nucleus A.

In deriving equation (19.19) we have assumed that just one target nucleus A is exposed to the beam. If there are ν nuclei A per unit area of the target and the cross sectional area of the beam is s we shall have for the total reaction rate

$$\begin{aligned} R &= \sigma j \nu s \\ &= \sigma I \nu \end{aligned} \tag{19.20}$$

where $I = sj$ is the beam current (we are assuming that the target is wider than the beam).

This expression for the reaction rate R is derived, of course, for a laboratory experiment in which a fixed target is bombarded by a beam. In Appendix E we derive an expression for the reaction rate in a totally different situation: that where the reacting nuclides consist of gases that are completely mixed with each other, as in a star or a thermonuclear reactor. Again the relevant parameter of the reacting nuclei is the cross section σ, and in fact it may be regarded as a characteristic of the reacting pair†, playing a role comparable with the half-life (or rather the decay constant) in the case of a radioactive nuclide. However, there is the difference that the cross section is not an absolute constant for the given pair, but can vary very strongly with energy, especially at the low energies prevailing in stellar interiors and nuclear reactors.

Recalling now from equation (19.14) that more than one reaction mode is usually possible for a given reacting pair, we see that we can define a cross section for each possible mode. Summing these cross sections over all possible modes, including elastic scattering, gives us the *total cross section* σ_{tot}; if we exclude the elastic scattering cross section σ_{el} from this summation we get the *total reaction cross section* σ_{react}, i.e.

$$\sigma_{\mathrm{tot}} = \sigma_{\mathrm{react}} + \sigma_{\mathrm{el}}. \qquad (19.21)$$

The significance of σ_{tot} is that it determines the rate of removal of projectiles from the beam in a laboratory experiment, and in fact it is usually determined by measuring the attenuation of the beam as it passes through the target. On the other hand, σ_{react} will determine the rate of depletion of the reacting nuclei, a quantity of considerable astrophysical interest; it also imposes an upper limit on the rate of formation of the compound nucleus, the balance corresponding to direct reactions (§21).

One might expect that σ_{react} would be of approximately the same order of magnitude as the geometrical cross section πR^2, although it turns out that it can be very much larger, and in any case is energy dependent. Nevertheless, with the nuclear radius R lying between 1 and 10 fm, a convenient unit in terms of which cross sections can be expressed is the *barn*, equal to 100 $\mathrm{fm}^2 (10^{-28}\ \mathrm{m}^2)$.

Finally, we note that one is often interested not only in the rate at which a given reaction proceeds but also in the angular distribution of the reaction products: these can emerge in all directions with respect to the beam‡ but

† The fact that the cross section is characteristic of the reacting *pair* means that its value remains unchanged if we exchange the roles of target and projectile, provided the CM energy is the same.

‡ In the CM frame of reference.

some directions will be more probable than others. Considering a two-body reaction of the form (19.10), we count the particles b emerging from the target with directions lying between θ and $\theta + \mathrm{d}\theta$ with respect to the beam (provided the incident particles and the target nuclei are unpolarised we shall have cylindrical symmetry with respect to the beam direction, and hence there will be no azimuthal dependence). We then write the number of such particles emerging per unit time from the target with the specified directions as

$$\mathrm{d}R(\theta) = \sigma(\theta) I v \, \mathrm{d}\Omega \tag{19.22}$$

where

$$\mathrm{d}\Omega = 2\pi \sin\theta \, \mathrm{d}\theta \tag{19.23}$$

is the solid angle defined by the directions θ and $\theta + \mathrm{d}\theta$. Then the total rate of this reaction is

$$R = \int_{\theta=0}^{\pi} \mathrm{d}R(\theta)$$

$$= Iv \int_{4\pi} \sigma(\theta) \, \mathrm{d}\Omega. \tag{19.24}$$

Then from equation (19.20) we see that

$$\sigma = \int_{4\pi} \sigma(\theta) \, \mathrm{d}\Omega. \tag{19.25}$$

The quantity $\sigma(\theta)$, which gives effectively the reaction rate per unit solid angle, for the direction θ, is known as the *differential cross section* for the reaction in question. We then refer to the quantity σ appearing on the left-hand side of (19.25), and in (19.20), as the *integrated cross section.*

Since all theories of nuclear reactions are formulated most naturally in the frame of reference where the centre of mass of the entire system is at rest, it will be necessary to relate the angle θ defined in this frame of reference to the corresponding angle in the laboratory-fixed system, this being the one that is actually measured, of course. This transformation is presented in Appendix A.

§20 GROSS ENERGY DEPENDENCE OF THE CROSS SECTION

Probably the most conspicuous features of the *excitation function* $\sigma(E)$ of nuclear reactions, i.e. of the energy dependence of their cross sections, are the very strong and rapid variations associated with resonances. These are discussed in the next section, and here we shall consider only the smoothed cross section, averaged over such structure. That is to say, we are concerned in this section only with gross trends in the energy independence of the cross section.

Beginning with neutron-induced reactions, where there is no complication coming from the Coulomb barrier, we might expect that the reaction cross section would be energy dependent, being simply equal to the geometrical cross section

$$\sigma_{\text{react}} = \pi R^2. \tag{20.1}$$

Experimentally it is found that this does hold fairly well at energies of 10 MeV or more, although there is a very slow decrease with increasing energy, suggesting a partial, and increasing, transparency of the nucleus. However, at lower energies the smoothed cross section increases much more rapidly with decreasing energy, varying according to

$$\begin{aligned} \sigma_{\text{react}} &\propto E^{-1/2} \\ &\propto v^{-1} \end{aligned} \tag{20.2}$$

at energies of a few electron volts or less, v being the neutron velocity before any interaction has taken place. The fact that the reactivity of neutrons increases as their energy decreases, a discovery made by Fermi and his collaborators in 1934, is of the utmost significance for nuclear reactors (see Chapter V).

The failure of the simple relation (20.1) at lower energies can easily be understood on quantum-mechanical grounds. At 10 MeV the de Broglie wavelength of a neutron is about 8 fm, which is comparable with the size of the target nucleus, so that the classical picture on which (20.1) is based should be fairly reliable. However, the de Broglie wavelength of a 1 eV neutron is about 2×10^4 fm, for which the concept of a well defined nuclear size breaks down completely. On the other hand, as the energy increases beyond 100 MeV, the wavelength will decrease to values not very much larger than the internucleon spacing. This accounts for the apparent transparency of the nucleus to high-energy neutrons, since it will be possible for them to pass right through without any significant interaction.

As for charged-particle projectiles, we may expect them to behave like neutrons at sufficiently high energies. However, at low energies their behaviour will be quite different: because of the Coulomb barrier that has to be penetrated before the two nuclei can begin to interact through their nuclear forces we may expect a rapid increase in reactivity with *increasing* energy.

We regard the low-energy situation as being represented by figure 20.1, which is very similar to figure 18.1, presented in connection with charged-particle *emission*. The only difference between the two cases is that now nuclei are approaching each other, of course, rather than separating. Assuming that reaction takes place once the deep well corresponding to the strongly attractive nuclear forces has been reached, we may write the cross section for both charged-particle and neutron reactions in the form

$$\sigma_{\text{react}} = a(E)T(E) \tag{20.3}$$

where $a(E)$ is an effective area that one nucleus presents to the other, and $T(E)$ represents the transmission coefficient of the barrier. Of course, in the case of neutron reactions there will be no barrier, just the deep, attractive well.

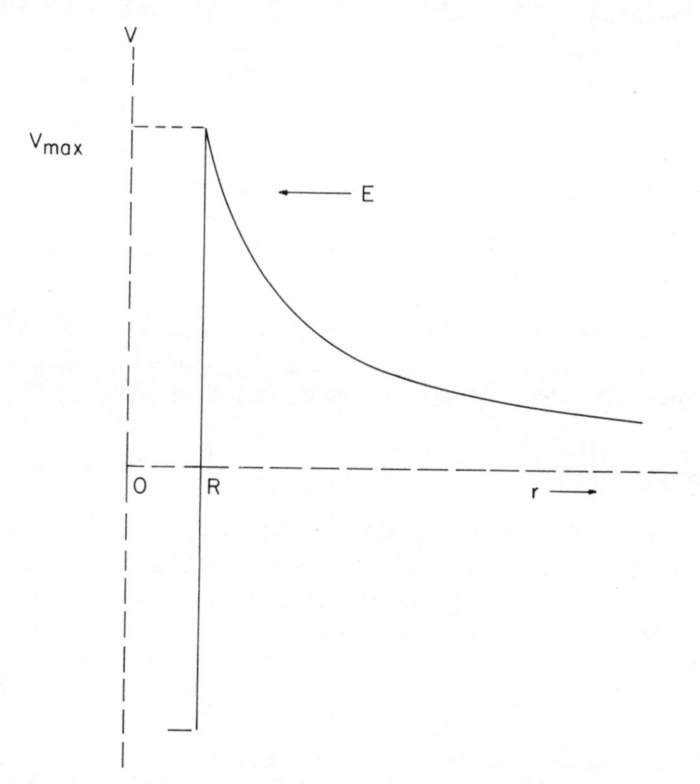

Figure 20.1 Penetration of the Coulomb barrier by bombarding projectiles.

To calculate the former quantity we note that an incident particle of momentum p (in the system where the centre of mass is at rest), and impact parameter b_l (§2) will have orbital angular momentum

$$l\hbar = b_l p. \tag{20.4}$$

Now in quantum mechanics l can take only integral values, while b_l cannot be measured exactly. Thus it is plausible to assert that all particles with orbital angular momentum $l\hbar$ will have impact parameters between b_l and b_{l+1}. The area of the associated annulus will then constitute an upper limit on the reaction cross section for particles of this angular momentum

$$\begin{aligned}\sigma^l_{\text{react}} &\leqslant \pi(b^2_{l+1} - b^2_l)\\ &= (2l+1)\pi\lambda\!\!\!^{-2}\end{aligned} \tag{20.5}$$

where $\lambda\!\!\!^- = \hbar/p$ is the de Broglie wavelength of relative motion. A more rigorous derivation of this result is presented in Appendix F.

The maximum possible value of l that can contribute to the reaction is determined by the nuclear radius R

$$l_{\max} = \frac{R}{\lambda\!\!\!^-}. \tag{20.6}$$

Then summing over all possible values of l we have

$$\sigma_{\text{react}} \leqslant \pi\lambda\!\!\!^{-2} \sum_{l=0}^{l_{\max}} (2l+1) = \pi\lambda\!\!\!^{-2}(l_{\max}+1)^2 = \pi(R+\lambda\!\!\!^-)^2. \tag{20.7}$$

Now this upper limit can be interpreted as the effective area $a(E)$, since the transmission coefficient $T(E)$ appearing in equation (20.3) can never exceed unity. For high energies, $\lambda\!\!\!^- \ll R$ we then have

$$a(E) = \pi R^2 \tag{20.8}$$

and for low energies, $\lambda\!\!\!^- \gg R$

$$a(E) = \pi\lambda\!\!\!^{-2}. \tag{20.9}$$

For high energies it is reasonable to assume a transmission coefficient of close to unity, in which case equation (20.3) reduces to equation (20.1). As for low energies, we combine (20.3) and (20.9) to give

$$\sigma_{\text{react}} \propto \frac{1}{E}\, T(E) \tag{20.10}$$

for which we shall calculate $T(E)$ in the following.

We deal first with charged particles, for bombarding energies considerably lower than the height of the Coulomb barrier

$$V_{\max} = \frac{Z_a Z_A e^2}{R_a + R_A} \simeq 1.25\,\frac{Z_a Z_A}{A_a^{1/3} + A_A^{1/3}}\ \text{MeV} \tag{20.11}$$

where we have used equation (4.2), with $r_0 = 1.15$ fm. The transmission coefficient for this case is given by the Gamow factor

$$T = C\exp(-2\pi n) \tag{20.12}$$

as in equation (18.5), C being the constant of proportionality appearing there. The Sommerfeld parameter, given by equation (18.9), is now expressed as

$$2\pi n = \gamma Z_a Z_A \sqrt{\frac{A}{E}} \qquad (20.13)$$

in which E represents the bombardment energy in the CM system, γ is still 0.99 $(\text{MeV})^{1/2}$, and A is again the 'reduced mass number' of equation (18.8). Then for low-energy charged-particle reactions we shall have from (20.10)

$$\sigma_{\text{react}} = \frac{S}{E} \exp(-2\pi n) \qquad (20.14)$$

where the S factor is, within the limits of this discussion, a constant, the numerical value of which is, of course, specific to the reacting pair of nuclei†.

Experimentally, it is found that the expression (20.14) does hold very roughly for charged-particle reactions at energies below the Coulomb barrier, provided once more the cross section is averaged over the rapid variations associated with resonances. Of course, an expression of the form (20.14) can always be fitted to the data, provided the S factor is allowed to be energy dependent: the test of the theory is then the extent to which S remains constant. As an example, consider the specific case of $^{16}O + {}^{16}O$, for which the S factor is found to decrease by a factor of about 100 over the range 7 to 12 MeV (in the CM system). This is a long way from constancy, but when we take note of the fact that the cross section itself increases by a factor of 10^5 over this same range we see that the parametrisation (20.14) does indeed take account of a large part of the energy dependence (actually, in this case the factor $\exp(-2\pi n)$ *over-compensates* the energy variation of the cross section).

The parametrisation (20.14) is particularly useful when we wish to extrapolate experimental cross sections to lower energies of astrophysical interest, for which the cross sections may be too small to be measured in the laboratory: the empirical extrapolation of the more slowly varying S factor will be inherently more reliable. It is for this reason that despite the extreme simplicity of the theory leading to equation (20.14), and the availability of more sophisticated theories, the cross sections of low-energy charged-particle reactions are still expressed in terms of the S factor defined as in equation (20.14).

† Note that the energy dependence of the $1/E$ factor in equation (20.14) is very weak compared with that of the Gamow factor, so that σ indeed vanishes for very low energies.

We stress that equation (20.14) is valid only for energies well below the Coulomb barrier. When the bombarding energy is much greater than the Coulomb barrier we may expect the geometrical result (20.1) to hold once more, with a gradual decrease again as the energy is increased still further.

In the case of neutrons, where there is no Coulomb barrier, the first interaction which the incident neutron will experience with the target nucleus will be the deep attractive potential shown in figure 20.1. A considerable reflection of the wavefunction of the incident neutron will occur at this sudden change of potential, so that here too the transmission into the internal region will certainly be less than 100%.

To analyse this further, we suppose that the deep attractive potential has the constant depth $V_0 = \hbar^2 K_0^2/2MA$, at least over a certain distance into the nuclear interior. Thus, while the incident neutron's wavefunction outside the interaction will have the form

$$\psi = a\mathrm{e}^{-\mathrm{i}kr} + b\mathrm{e}^{\mathrm{i}kr} \qquad (r > R) \tag{20.15}$$

where $k = (2MAE)^{1/2}/\hbar$, inside it will have the form

$$\psi = c\mathrm{e}^{-\mathrm{i}Kr} \qquad (r < R) \tag{20.16}$$

where

$$K^2 = k^2 + K_0^2. \tag{20.17}$$

(Note that because we are at low energy, we consider just the s wave i.e. $l = 0$.)

It is to be particularly noted that we are assuming that the incident neutron wave is completely absorbed as it propagates towards the centre of the nucleus, so that in (20.16) there is no wave $\mathrm{e}^{\mathrm{i}Kr}$ returning from the nuclear interior. This means that we do not regard the potential $V = -V_0$ as being defined throughout the nuclear interior: the interaction between the incident neutron and the target nucleus is too complicated to be represented simply as a function of the position of the former.

The transmission coefficient for the jump in potential V_0 at the surface of the interaction region is given by

$$T = \left|\frac{c}{a}\right|^2 \frac{K}{k}. \tag{20.18}$$

Applying now the usual continuity conditions on ψ and its first derivative determines c/a, from which (see Problem IV.4)

$$T = 4\frac{k}{K_0} \tag{20.19}$$

if the incident energy is low compared with the well depth V_0, i.e. if $k \ll K_0$. Then, with $k \propto E^{1/2}$, equation (20.10) will reduce to the $1/v$ law (20.2)†.

Optical model. The foregoing treatment provides, in a relatively simple way, a qualitative understanding of the energy dependence of smoothed reaction cross sections at the opposite limits of high and low energy. Quantitatively, however, it is quite inadequate, and it leaves untreated altogether the cross-over region of intermediate energy.

If we are to obtain any quantitative improvement, and have a coherent approach valid at all energies, it will be necessary to adopt a much more thorough quantum-mechanical treatment of the reaction problem, and in particular avoid introducing the rather nebulously defined effective area $a(E)$. This means that we shall have to solve a Schrödinger equation

$$H\Psi = E\Psi \tag{20.20}$$

where

$$H = \hat{T} + \hat{V} \tag{20.21}$$

in which $\hat{T}$ is the kinetic energy operator and $\hat{V}$ the potential energy operator, as in equations (5.2)–(5.5).

In an exact treatment of the reaction problem we should have to take account of each nucleon explicitly. Thus for the potential energy we would have

$$\hat{V} = \sum_{i>j}^{A_a+A_A} \sum_{j=1}^{A_a+A_A} V_{ij} \tag{20.22}$$

where the summation goes over the interaction potential V_{ij} between each pair of nucleons in the *two* nuclei A and a. The kinetic energy operator $\hat{T}$ would likewise involve a summation of the corresponding operators for each nucleon in the two nuclei, while Ψ would be a function of all the nucleons of the two nuclei: $\Psi = \Psi(r_1, r_2, \ldots, r_{A_a}, \ldots, r_{A_a+A_A})$

Since exact solution of the many-body Schrödinger equation even for a single nucleus is, as we indicated in §5, an impossibility, we cannot hope to solve the exact Schrödinger equation for *two* interacting nuclei. However, recalling that in the case of the bound states of a single nucleus it was possible to construct an approximating model, the shell model, for which the Schrödinger equation *was* soluble, it is reasonable to expect that a similar approach might also be possible in the present case.

Following the spirit of the shell model, our reaction model consists of averaging completely over the internal structure of each nucleus and

† A simple way of understanding this law is to argue that the probability that a neutron reacts with the target nucleus is simply proportional to the time that the neutron spends in the vicinity of the nucleus.

representing the interaction between them simply as a function of the distance r between the two centres, $V = \mathcal{V}(r)$. The Schrödinger equation (20.20) then reduces to an equation in the single vector variable $\boldsymbol{r}$.

$$H_0\psi(\boldsymbol{r}) = E\psi(\boldsymbol{r}) \tag{20.23}$$

where

$$H_0 = -\frac{\hbar^2}{2AM}\nabla^2 + \mathcal{V}(\boldsymbol{r}) \tag{20.24}$$

the first term of which is the kinetic energy operator for the relative motion of the two nuclei, A denoting $A_a A_A/(A_a + A_A)$ again, and ∇^2 the Laplacian corresponding to $\boldsymbol{r}$.

The one-body Schrödinger equation (20.23) is certainly soluble, even though for most forms of potential $\mathcal{V}(\boldsymbol{r})$ a computer will be necessary. The important question now is: can it have any relevance to the reaction processes taking place between the nuclei A and a?

Clearly, since the only variable appearing in the Schrödinger equation (20.23) is the separation $\boldsymbol{r}$ of the 'entrance channel' nuclei A and a, there is no possibility of *explicitly* representing reactions in which other nuclei are formed. Thus it might appear that the only process that could be represented in this way would be elastic scattering. However, if the potential $\mathcal{V}(\boldsymbol{r})$ is assumed to be complex we shall see that there can be an attenuation of the wavefunction $\psi(\boldsymbol{r})$ associated with the relative motion of the entrance channel nuclei A and a. But such a loss of flux in the entrance channel is precisely what happens when reactions are taking place, so that the model corresponding to equation (20.23) can be made to yield, in addition to the differential elastic scattering cross section, $\sigma_{\text{el}}(\theta)$, the total reaction cross section, σ_{react}, although we shall not, of course, be able to calculate the cross section for particular reactions in this way.

To see how a complex potential $\mathcal{V}(\boldsymbol{r})$ can indeed give rise to absorption, consider a one-dimensional problem in which the complex potential is a square well, width x_0

$$\mathcal{V}(x) = \begin{cases} -(V_0 + \mathrm{i}W_0) & (-\frac{1}{2}x_0 < x < \frac{1}{2}x_0) \\ 0 & (x > \frac{1}{2}x_0,\ x < -\frac{1}{2}x_0). \end{cases} \tag{20.25}$$

The solution to the Schrödinger equation (20.23) inside this well takes the form

$$\psi(x) = a\mathrm{e}^{-\mathrm{i}\varkappa x} + b\mathrm{e}^{\mathrm{i}\varkappa x} \qquad -\tfrac{1}{2}x_0 < x < \tfrac{1}{2}x_0 \tag{20.26}$$

where

$$\begin{aligned} \varkappa^2 &= \frac{2MA}{\hbar^2}(E + V_0 + \mathrm{i}W_0) \\ &= K_0^2 + k^2 + \mathrm{i}q^2 \end{aligned} \tag{20.27}$$

in which

$$K_0^2 = \frac{2MA}{\hbar^2} V_0 \tag{20.28a}$$

$$k^2 = \frac{2MA}{\hbar^2} E \tag{20.28b}$$

$$q^2 = \frac{2MA}{\hbar^2} W_0. \tag{20.28c}$$

Writing

$$\varkappa = K + \mathrm{i}\gamma \tag{20.29}$$

where K and γ are real, we shall have from (20.27)

$$K^2 - \gamma^2 = K_0^2 + k^2 \tag{20.30a}$$

$$2K\gamma = q^2. \tag{20.30b}$$

We shall solve this pair of equations for K and γ below, but it suffices for the moment to note from (20.30b) that K and γ must have the same sign, and we can choose both to be positive. Then writing (20.26) as

$$\psi = a\mathrm{e}^{\gamma x}\mathrm{e}^{-\mathrm{i}Kx} + b\mathrm{e}^{-\gamma x}\mathrm{e}^{-\mathrm{i}Kx} \qquad -\tfrac{1}{2}x_0 < x < \tfrac{1}{2}x_0 \tag{20.31}$$

we note that each term represents progressive waves whose amplitudes decrease in the direction of propagation, i.e. the wave is absorbed as it traverses the medium.

The solutions to equations (20.30) are

$$K = \{\tfrac{1}{2}[\sqrt{(k^2 + K_0^2)^2 + q^4} + k^2 + K_0^2]\}^{1/2} \simeq (k^2 + K_0^2)^{1/2} \tag{20.32a}$$

$$\gamma = \{\tfrac{1}{2}[\sqrt{(k^2 + K_0^2)^2 + q^4} - k^2 - K_0^2]\}^{1/2} \simeq \tfrac{1}{2}q^2/K \tag{20.32b}$$

the approximate forms holding for sufficiently weak absorption, $q \ll K_0$. In this latter case, noting that K is the wavenumber in the well, we see that the flux density will be attenuated by a factor e over $V_0/(2\pi W_0)$ wavelengths, provided also the total energy E (i.e. the kinetic energy outside the well) is small compared with the well depth V_0, i.e. $k \ll K_0$.

The behaviour of the wavefunction in this complex potential is reminiscent of the passage of light through a material medium, where there is not only refraction but also absorption. For this reason the complex potential $\mathscr{V}(r)$ of nuclear physics is known as the *optical potential*.

However, the square-well form that we discussed above has too sharp a discontinuity at the edge to be realistic, and in practice one prefers the Saxon–Woods form (12.21) used for the shell-model potential:

$$\mathscr{V}(r) = -\frac{V_0 + \mathrm{i}W_0}{1 + \exp[(r-R)/a]}. \tag{20.33}$$

The choice of the same form of potential as for the shell model is obvious when one of the two nuclei is a single nucleon, but it is also often used for the more general case of nucleus–nucleus reactions. Further refinements consist of allowing the parameters R and a to take different values for the real and imaginary terms. Another possibility for the imaginary term is to replace the Saxon–Woods form by a Gaussian peaked in the surface. When the incident nucleus has spin, and particularly when it is a single nucleon, it is customary to include a spin–orbit term, of the same form as that adopted in the shell-model potential, equation (12.22). More complicated kinds of spin dependence have also been considered. Finally, it is necessary to add a Coulomb potential for all projectiles other than neutrons.

The actual way in which the total reaction cross section σ_{react} and the differential elastic cross section $\sigma_{\text{el}}(\theta)$ are determined for a given complex potential $\mathscr{V}(r)$ is to solve the Schrödinger equation (20.23) with the wavefunction $\psi(r)$ satisfying the boundary condition appropriate to the reaction experiment

$$\psi(r) \sim \mathrm{e}^{ikz} + f(\theta)\,\frac{\mathrm{e}^{ikr}}{r} \tag{20.34}$$

the first term of which represents the incident plane wave of the beam and the second the outgoing spherical scattered wave (see Appendix F). Proceeding by trial and error it is possible to adjust the parameters of the optical potential to get essentially exact fits to the angular distribution $\sigma_{\text{el}}(\theta)$ of the elastic scattering and to the reaction cross section σ_{react} at any one energy. However, when data at different energies are considered it is found that the parameters must be allowed to be energy dependent.

Considerable theoretical progress in relating the optical model to the basic nucleon–nucleon forces has been made in recent years, both for the nucleon–nucleus and nucleus–nucleus cases. While a complete quantitative success remains elusive there is at least a considerable qualitative understanding of many of the features of the optical model, including the energy dependence of the parameters. This means that while the use of the optical model to extrapolate charged-projectile reactions to very low energies is not completely unambiguous, it is certainly more reliable than the simple Gamow-factor approach.

To conclude this discussion of the optical model we point out that in addition to its inability to distinguish between different types of reaction there is a further limitation. When the entire interaction (20.22) between two nuclei is smoothed out and replaced by a simple potential function $\mathscr{V}(r)$, it is impossible to represent the brusque variations of cross section associated with resonances, these originating in excitations of the internal structure of the interacting nuclei. However, this is not a serious defect in the present discussion where we have been concerned only with average trends in the cross section.

§21 THE COMPOUND NUCLEUS

The optical model, discussed at the end of the previous section, describes in general terms the interaction between the entrance channel nuclei A and a, i.e. the target and the projectile. In particular, the model indicates whether they simply scatter elastically off each other (and if so, with what angular distribution), or whether they actually initiate a reaction, in the sense that a net transfer of nucleons between the two nuclei takes place, or at the very least one or other of the nuclei is excited (inelastic scattering). The actual mechanism by which elastic scattering takes place is fairly easy to visualise: it is simply the result of the net interaction between the two nuclei, averaged over their internal structure. However, nothing has been said so far about the mechanism leading to reactions.

It turns out that a significant clue to the nature of reaction mechanisms

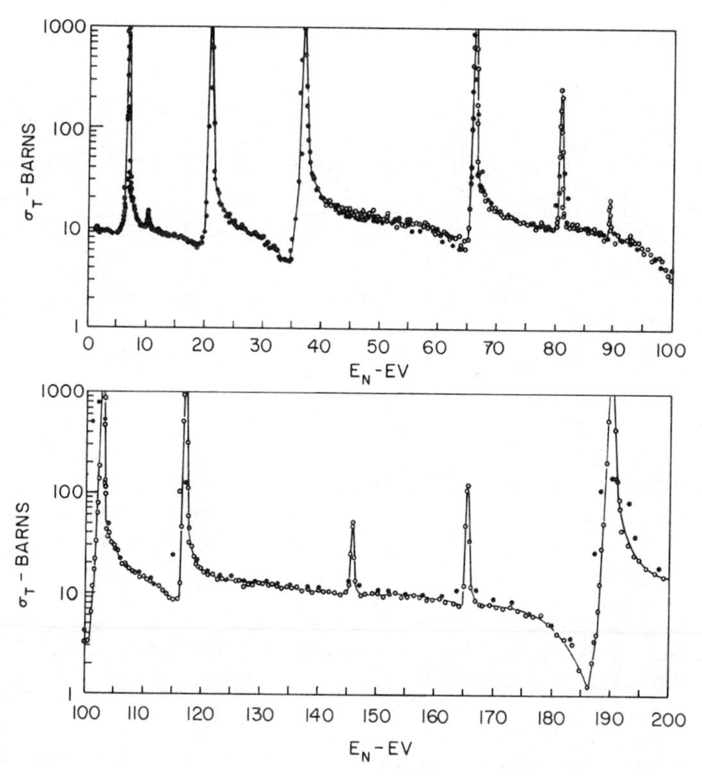

Figure 21.1 Total cross sections for neutrons on ^{238}U. Adapted from *Brookhaven National Laboratory Report* BNL 325, 2nd edn, Suppl. 2 (1965).

is provided by the phenomenon of *resonance* to which we have alluded in the previous section. These rapid variations of cross section with energy constitute one of the most striking phenomena in the whole of nuclear physics, after fission. This behaviour is manifested most clearly in reactions induced by low-energy neutrons, and it was in these reactions that the phenomenon was discovered in the period 1935–6†. It was found that the cross section could vary by factors of thousands over an energy interval as small as 1 eV in the case of heavy nuclei. We show in figures 21.1 and 21.2 some more modern results for neutron cross sections; the resonances are seen to be broader and more widely separated for lighter targets.

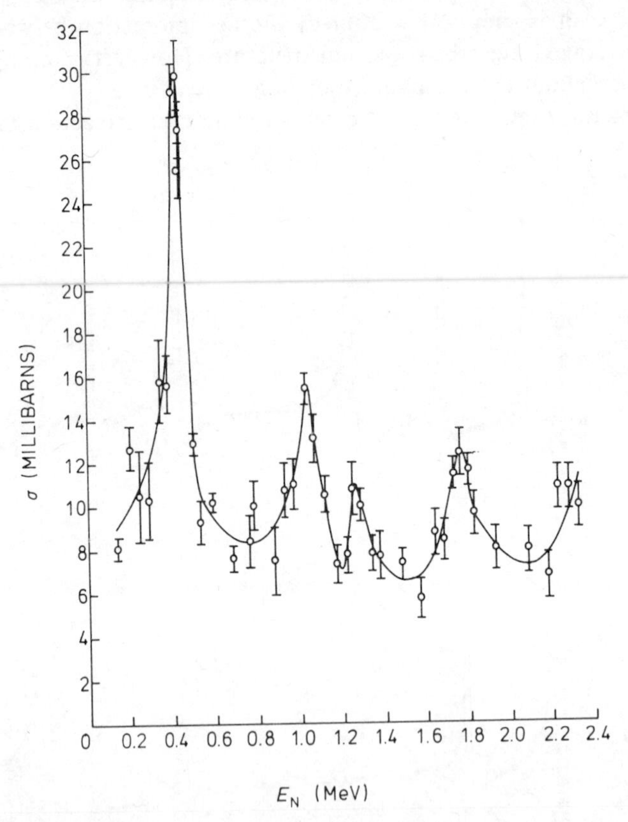

Figure 21.2 Cross sections for (n,γ) reactions on ^{11}B. Adapted from *Brookhaven National Laboratory Report* BNL 325, 2nd edn, Suppl. 2 (1965).

† For an excellent review of the discovery of resonances and the whole question of the compound nucleus see the article by F L Friedman and V F Weisskopf 1955 in *Niels Bohr and the Development of Physics*, edited by W Pauli (New York: McGraw-Hill).

Resonances are also found at low energies for reactions induced by light charged particles, such as protons and α particles, as seen for example in figures 21.3, 21.4 and 21.5. The distorting effect of the Coulomb potential is particularly striking in the first case, where the bombarding energy is well below the Coulomb barrier.

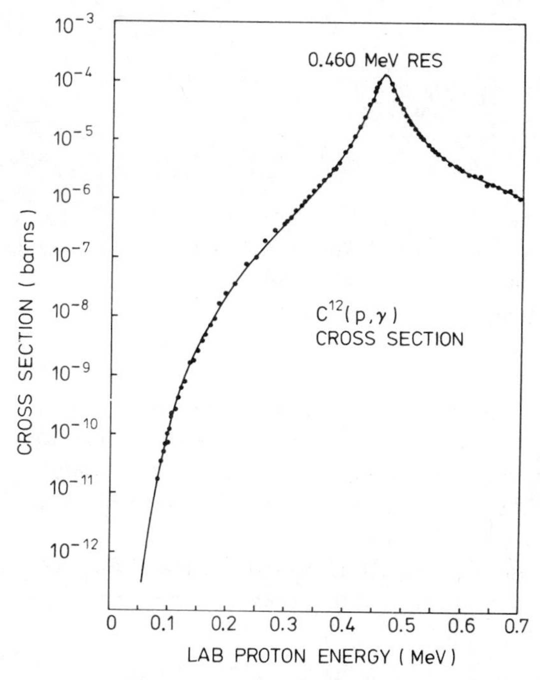

Figure 21.3 Cross section for (p, γ) reaction on ^{12}C. Adapted, with permission, from Fowler W A, Caughlan G R and Zimmerman 1967 *Ann. Rev. Astron. Astrophys.* **5** 525 © 1967 by Annual Reviews Inc.

For reasons that will be discussed below, the resonances in all these reactions, both those induced by neutrons and those induced by light charged particles, become broader and at the same time more closely spaced as the bombarding energy increases, until finally they overlap completely, all structure then disappearing from the excitation curve (see figure 21.6). Nevertheless, at these same relatively high energies, a new type of much wider resonant structure is often seen in heavy-ion reactions, as shown for example in figure 21.7. (Note that because of the Coulomb barrier these reactions have negligible cross section at lower energies, except of course for the Rutherford contribution to elastic scattering.)

We now discuss the implications of resonances for the reaction mechanism, remarking first that since they constitute a discrete set, they are

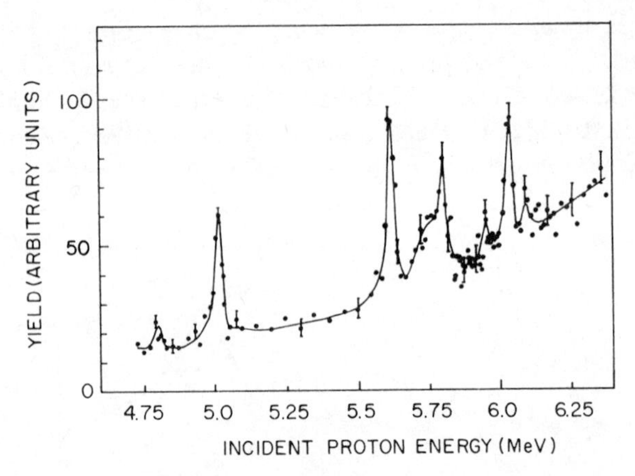

Figure 21.4 Cross section for inelastic scattering of protons on ^{89}Y (excitation of state at 0.92 MeV). Adapted from Jones G A, Lane A M and Morrison G C 1964 *Phys. Lett.* **11** 329.

somewhat reminiscent of bound states. However, it is obvious that the states of the combined system of the two reacting nuclei to which they correspond are not truly bound, and furthermore we see that they do not occur at precise energies, but rather over a finite interval corresponding to the width Γ of the resonant peak.

We may interpret this width Γ as an *uncertainty* in the energy of the resonant states. At the same time, since these states are not bound, we shall have

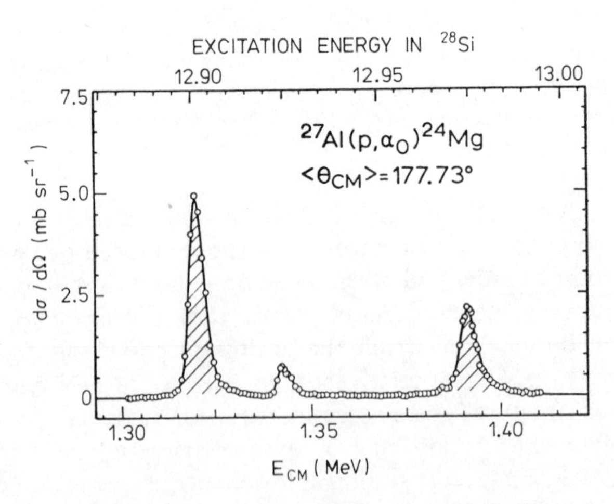

Figure 21.5 Differential cross sections ($\theta \simeq 180°$) for ^{27}Al (p, α) ^{24}Mg reaction. Adapted from Driller H, Blanke E, Genz H, Richter A, Schrieder G and Pearson J M 1979 *Nucl. Phys.* A **317** 300.

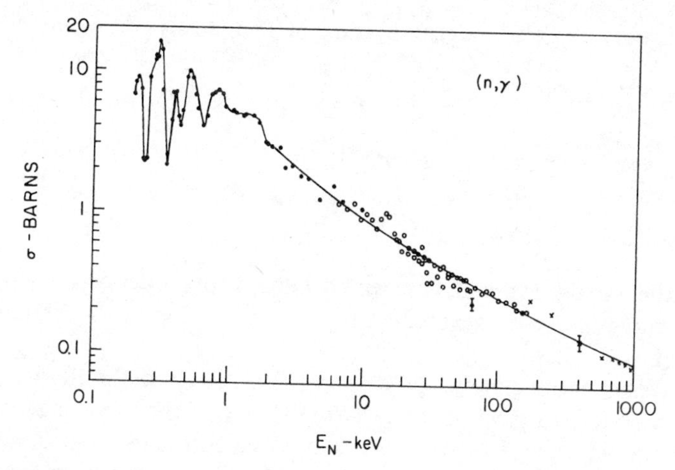

Figure 21.6 Integrated cross sections for radiative capture of neutrons on natural Pt. Adapted from *Brookhaven National Laboratory Report* BNL 325, 2nd edn, Suppl. 2 (1965).

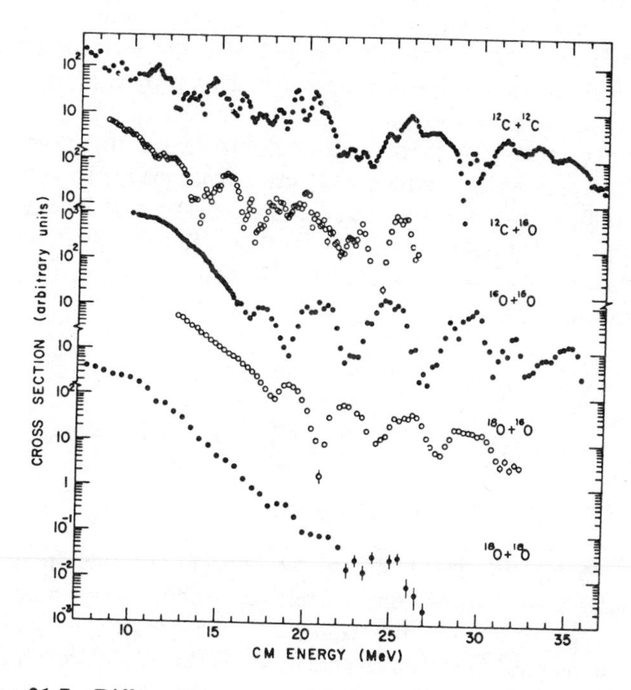

Figure 21.7 Differential cross section ($\theta = 90°$) for elastic scattering in various heavy-ion systems. Taken with permission from Siemssen R H 1976 *Nuclear Spectroscopy and Reactions*, ed J Cerny, ch. IV.C.1. (New York: Academic)

to associate with them a decay lifetime τ. The Heisenberg principle, in the form

$$\Delta E\, \Delta t \sim \hbar \tag{21.1}$$

relates these two quantities according to

$$\tau = \frac{\hbar}{\Gamma} = \frac{6.6 \times 10^{-16}}{\Gamma(\text{eV})}\ \text{s}. \tag{21.2}$$

Hence the narrowest resonances with a width of the order of 1 eV, occurring in low-energy neutron reactions on heavy nuclei, imply a lifetime of the order of 10^{-15} s.

This result immediately gives us some insight into the nature of these resonant states. In view of the general validity of the independent-particle model, the simplest picture of these states that comes to mind is that the incident neutron is trapped in the shell-model potential (to which the real part of the optical potential obviously bears a close relation) and moves to and fro across the nucleus without colliding with any of the nucleons of the target nucleus. Now we have seen (p. 59) that the depth V_0 of the shell-model potential is about 50 MeV. But since the projectile neutron can escape with an energy equal to the bombarding energy E (in the CM system), it follows that its kinetic energy inside the shell-model potential must be at least 50 MeV. Thus the time taken by a projectile neutron of low incident energy to cross the nucleus is of the order of 10^{-22} s.

Of course, even though there is no Coulomb barrier to be penetrated by the neutron, we should not expect it to escape from the nucleus the first time it reaches the surface, since a considerable reflection of its wavefunction occurs at the sudden change in potential there. Rather, the lifetime of the resonant state in this picture should be of the order of $T^{-1} \times 10^{-22}$ s, where T is the transmission coefficient given by equation (20.19):

$$T = \frac{4k}{K_0} = 4\sqrt{\frac{E}{V_0}}\,. \tag{21.3}$$

Thus we obtain finally for the lifetime of the resonant state in this picture

$$\tau = \frac{10^{-19}}{\sqrt{E(\text{eV})}}\ \text{s}. \tag{21.4}$$

Now the narrow low-energy neutron resonances observed in heavy nuclei have lifetimes thousands of times longer than is implied by equation (21.4), so that they cannot possibly be described by this simple picture. Rather we must invoke the mechanism conceived by Niels Bohr in 1936, which we now describe. We shall do so, however, in a more modern framework, retaining in particular the idea of the shell-model potential, which did not appear in Bohr's formulation,

Still considering for simplicity the case where the bombarding projectile is a single nucleon, we must suppose that while it will certainly feel a mean overall potential, it will also be able to collide with individual nucleons in the target nucleus, and transfer energy to them. If the energy of the incident nucleon is sufficiently low, it may well be that after its first encounter with a target nucleon it loses enough energy for it to fall into one of the unoccupied bound states of the shell-model potential (figure 21.8), so that it will not be able to escape from the well. At the same time, the struck nucleon may not acquire sufficient energy to escape either, and so will simply be excited to one of the unoccupied bound states.

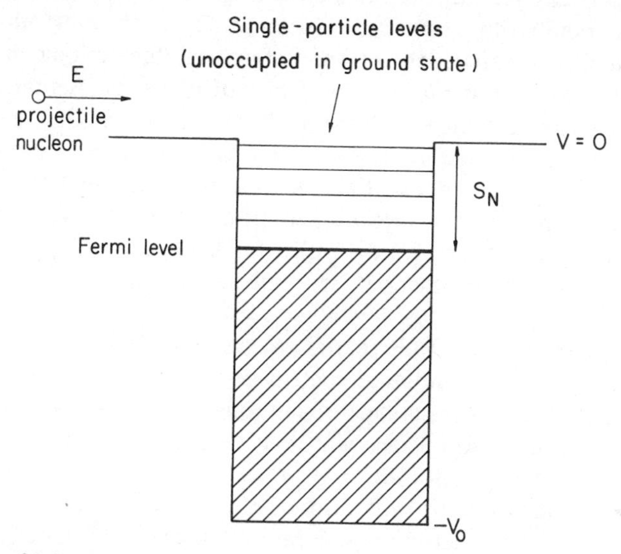

Figure 21.8 Shell-model picture of formation of compound nucleus with nucleon projectile.

Since these single-particle states are discrete, we see at once that this mechanism can likewise function only for certain particular values of the bombarding energy, which it is tempting to associate with the resonant energies. On the other hand, as the bombarding energy increases, there will be more and more combinations of single-particle states with energies lying in a given interval. Thus we can account quite naturally for the observation that the resonances become more closely spaced as the bombarding energy increases.

In any case, we now have the picture that after the first collision between the incident nucleon and a target nucleon, *all* nucleons will be trapped in bound states of the shell-model potential. It follows that both the incident nucleon and the struck target nucleon will undergo further collisions, both with each other and with the other target nucleons. In this way, the incident energy will eventually be shared among all the target nucleons, and a situation of statistical equilibrium established; we speak of formation of the *compound nucleus*.

Because the compound nucleus can emit the incident nucleon with an energy equal to the bombarding energy E (in the CM system) it follows that the compound nucleus must be formed in an excited state, the energy of excitation being $E + S_N$, where S_N is the nucleon separation energy from the ground state of the compound nucleus†. However, with this energy being shared between all nucleons in the compound nucleus, no nucleon (or group of nucleons) will have sufficient energy to escape from the equilibrium state, unless the bombarding energy E is large. Thus disintegration of the compound nucleus will depend on some statistical fluctuation concentrating sufficient energy on one nucleon, or group of nucleons. Actually, for low bombarding energies there will be very few decay modes energetically available to the compound nucleus. One mode that is always possible consists of the emission of a particle identical to the incident projectile, and with the same energy: this is elastic scattering, of course (we speak here of *compound-elastic* scattering, in contrast to elastic scattering that proceeds directly via the optical potential, without any formation of a compound nucleus).

The only other decay mode that is *always* possible for low bombarding energy consists of the compound nucleus getting rid of all its excess energy by emitting a photon (or cascade of photons) before any nucleons can be emitted: the compound nucleus will then be stable and the projectile will have been radiatively captured.

The compound nucleus will also be able to disintegrate into modes corresponding to exothermic reactions, but there are generally very few such possibilities, and maybe none at all for certain pairs of reacting nuclei. However, as the bombarding energy increases the compound nucleus will be able to decay into more and more endothermic modes, and as a result the lifetime of the compound nucleus will decrease. This is one reason why the width of the resonances increases with bombarding energy. (The other reason is that the rate of each decay mode will increase with energy, because of enhanced transmission of the separating particles.)

We now remark on the role of the shell model in this description of the compound-nucleus reaction mechansim. While we have drawn heavily on shell-model concepts, it must be realised that without deviations from the shell model this mechanism could not work. In the first place, if every nucleon in the target nucleus always remained in the same single-particle state, the energy of the incident nucleon or nucleus could not be shared among them and no compound nucleus would be formed. Secondly, even

† Separation energy is defined in §17.

if the compound nucleus could be formed, the emission of nucleons, or groups of nucleons, by the compound nucleus is completely forbidden in a strict shell-model picture, according to which each nucleon remains in the same single-particle state until an electromagnetic decay intervenes — radiative capture would be the only possible process besides elastic scattering.

It was for this reason that the immediate success of Bohr's introduction of the compound nucleus put an end, for the time being, to the speculations on the possible validity of a nuclear shell model that were current in the 1930s. In particular, the rapid sharing of the incident energy among all the target nucleons was held to be quite antithetical to the idea of independent-particle motion. A whole decade was to elapse before the shell model was revived, and even then only in the face of inescapable evidence.

The point is, that while the shell model undoubtedly has a large measure of validity, the deviations from it are sufficient to permit formation and decay of the compound nucleus. More specifically, we know that because of the Pauli principle the scattering of two nucleons bound in a nucleus is much weaker than the scattering of two free nucleons (§13). But during the energy-sharing process the interaction between the nucleons with an excess energy and the others will be close to the free interaction. Hence there is no incompatibility at all between the rapid energy-sharing associated with formation of the compound nucleus, and the validity of the shell-model description of bound nuclei.

On the other hand, once the compound nucleus has been formed and the incident energy shared amongst all the target nucleons, the action of the Pauli principle will be much more effective in damping further scatterings between nucleons. Thus we may expect the compound nucleus to be long lived, as compared with the time taken for its formation. All in all, we see that the shell model, far from being incompatible with the compound nucleus, actually enhances our understanding of it; in particular, we can reconcile the rapid formation of the compound nucleus with its relative stability once formed.

It must be emphasised that while this picture of the compound-nucleus mechanism was inspired by the observation of isolated resonances at low energy, there is no *a priori* reason why the same mechanism should not be operative at higher energies, where the absence of a visible resonant structure could be due simply to the greater width and closer spacing of the resonances. Actually, as the bombarding energy increases, a completely different reaction mechanism does become possible, as we shall now see.

In the foregoing description of the formation of the compound nucleus the importance of the first collision between the incident projectile and a target nucleon will be apparent. With increasing bombarding energy it becomes more and more likely that after the first collision either the projectile or the struck nucleon (or group of nucleons) will have sufficient energy to emerge†, so that the reaction will have taken place without formation of the compound nucleus. We say in this case that the reaction has been mediated by the *direct interaction* mechanism.

It turns out that for certain states of the final product nuclei the reaction proceeds mainly by direct interaction, with compound-nucleus formation playing a negligible role. But at the bombarding energies where this happens many different reaction modes are possible, and it is found generally that the overall reaction rate is still dominated by the compound-nucleus mechanism. Thus, important as direct reactions may be as a tool in nuclear spectroscopy, we confine ourselves henceforth to compound-nucleus processes, except to say that the resonant structure seen in heavy-ion reactions (figure 21.7) may be interpreted in terms of an intermediate mechanism lying between the two limits of compound-nucleus formation and direct interaction.

Independence hypothesis. Because of the extremely long lifetime of the compound nucleus (compared with the average time of transit of a nucleon across the nucleus), Bohr suggested that its mode of decay should be independent of its mode of formation, i.e. the compound nucleus 'forgets' how it was formed. (We would not expect this in the case of direct reactions, which, being single-stage processes, take place much more rapidly.) Of course, global conservation laws will have to be respected: the decaying compound nucleus must 'remember' to conserve angular momentum and parity, for example.

One of the most direct experimental implications of the independence hypothesis is that angular distributions in the CM system should be symmetric with respect to reflection along the beam direction, i.e.

$$\sigma(\theta) = \sigma(\pi - \theta). \tag{21.5}$$

This 'fore-aft' symmetry means essentially that the compound nucleus forgets by which of the two modes shown in figure 21.9 it was formed. One must avoid the temptation of concluding that this nuclear amnesia implies that angular distributions are isotropic. Angular momentum has to be conserved and the vector of this is of necessity confined to the plane perpendicular to the beam, although it will have a random orientation within this plane. Thus this plane will be a privileged one and different orientations with respect to it will be distinguishable; the two sides of it, however, will not be.

It must be realised that there are certain circumstances where this fore-aft symmetry will hold automatically, even if the reaction does not proceed by way of the compound-nucleus mechanism, e.g. if the target nuclei and pro-

† Another possibility is that the projectile picks up nucleons from the target nucleus, or deposits nucleons (if the projectile is complex) on the target nucleus.

jectile are identical. On the other hand, deviations from 'fore-aft' symmetry certainly imply a failure of the independence hypothesis. In fact, 'fore-aft' symmetry is often found not to hold exactly. Frequently, this is due to a direct interaction contribution, which tends to lead to a forward peaking, especially in elastic scattering.

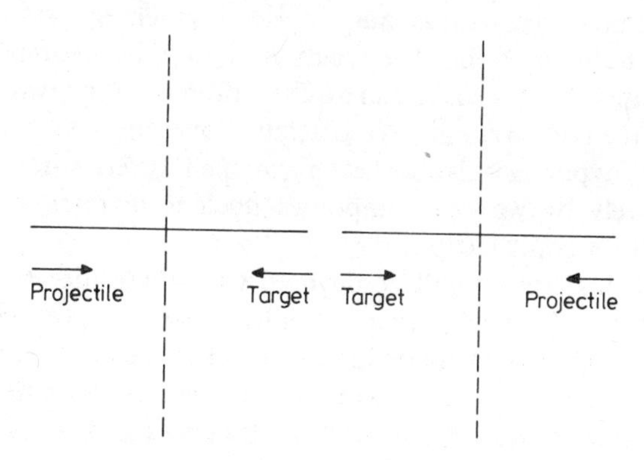

Figure 21.9 Independence hypothesis and 'fore-aft' symmetry.

We shall now examine more closely the relation between the compound nucleus mechanism and the independence hypothesis, and we shall see in fact that even when the reaction proceeds entirely by compound nucleus formation, there are situations where the validity of the independence hypothesis cannot be assured.

The independence hypothesis might be expected to hold best at low energies, where the lifetime of the resonant states is longest. Indeed, insofar as the resonances in this region are isolated, the long lifetime implying extreme narrowness, the compound nucleus will be formed in a definite quantum state, the decay properties of which must be independent of the mode of formation.

As the lifetime of the resonant states decreases with increasing bombarding energy, deviations from the independence hypothesis should be expected. More specifically, with the resonances beginning to overlap, the compound nucleus will effectively be formed in a mixture of quantum states, the decay properties of which will depend on the phase relation between the states in the mixture, which in turn will be determined by the mode of excitation of the compound nucleus.

However, as we go to still higher energies, where cross sections vary smoothly with energy because of the complete overlap of resonances, there is reason to believe that the independence hypothesis will be satisfied once more. This is because with so many resonant states contributing at any one energy the net effect associated with the relative phases may be expected to be very small, simply on the basis of randomisation: the 'interference terms' wash out†. For this reason this region is spoken of as the *statistical*, or *continuum*, region.

This discussion suffices to show that the independence hypothesis and the compound nucleus mechanism are by no means synonymous. Thus, although the independence hypothesis is satisfied both in the low-energy region of isolated resonances and in the high-energy statistical region (assuming in both cases that the reaction is entirely a compound nucleus one), the reasons for this success are quite different in the two regions. And in the intermediate region of partially overlapping resonances the independence hypothesis is not even satisfied, even when the reaction proceeds entirely by way of compound nucleus formation, without any direct interaction contribution.

The rest of this section will be devoted to the consideration of another property that must hold whenever the independence hypothesis is satisfied. This property, relating to angle-integrated cross sections rather than differential cross sections, is less striking than the fore-aft symmetry, but it has important consequences. We assert that the cross section of any reaction should be expressible as the product of a factor characteristic of the formation of the compound nucleus, and another factor characteristic of its decay, insofar as these two stages can be regarded as being independent of each other.

Let us denote the two reacting nuclei A and a by α (this labels the so-called *entrance channel*), and the product nuclei of each of the various reaction modes by β, β', etc (these label the so-called *exit channels*). Then with formation and disintegration of the compound nucleus being regarded as independent events, we can write for the integrated cross section of the reaction $\alpha \rightarrow \beta$

$$\sigma_{\alpha\beta} = \sigma(\alpha) P_\beta \tag{21.6}$$

where $\sigma(\alpha)$ is the cross section for formation of the compound nucleus from the entrance channel α and P_β is the probability that the compound nucleus decays into channel β. Clearly

$$\sum_\beta P_\beta = 1 \tag{21.7}$$

where the sum goes over all possible exit channels.

Actually, since the total angular momentum J and the parity π of the entire reacting system have to be conserved, the factorisation property must apply to each set of J and π values‡, thus

$$\sigma_{\alpha\beta} = \sum_{J,\pi} \sigma^{J\pi}(\alpha) P_\beta^{J\pi} \tag{21.8a}$$

† This qualitative argument can be justified theoretically.

‡ The generalisation to the case where there are other conserved quantities is straightforward.

where $\sigma^{J\pi}(\alpha)$ represents the cross section for formation of the compound nucleus in resonant states characterised by quantum numbers J and π, while $P_\beta^{J\pi}$ represents the probability of decay from such states into the channel β. In place of (21.7) we must now have for each (J, π)

$$\sum_\beta P_\beta^{J\pi} = 1. \tag{21.9}$$

We now combine this factorisation property with another, known as *detailed balance*. To understand this latter property we note first that it might be thought reasonable to assert that the cross section of any reaction $\alpha \to \beta$ should be equal to that of its inverse $\beta \to \alpha$, i.e. $\sigma_{\alpha\beta} = \sigma_{\beta\alpha}$, provided that the energies involved are the same in both cases. In fact, $\sigma_{\alpha\beta}$ and $\sigma_{\beta\alpha}$ each contain a statistical factor relating to the exit channel, β and α, respectively, and it may be shown that the equality for the two directions of the reaction applies not to the cross sections themselves, but rather to the cross sections without the statistical factors†. The detailed balance property is thus

$$\frac{\sigma_{\alpha\beta}}{k_\beta^2 g_\beta} = \frac{\sigma_{\beta\alpha}}{k_\alpha^2 g_\alpha} \tag{21.10}$$

where k_α and k_β are the respective wavenumbers in the α and β channels, while

$$g_\alpha = (2j_\mathrm{a} + 1)(2j_\mathrm{A} + 1), \text{ etc} \tag{21.11}$$

in which j_a is the spin of nucleus a, etc.

We now substitute (21.8a), and the corresponding relation for the inverse reaction,

$$\sigma_{\beta\alpha} = \sum_{J,\pi} \sigma^{J\pi}(\beta) P_\alpha^{J\pi} \tag{21.8b}$$

into (21.10), obtaining

$$\sum_{J,\pi} (k_\alpha^2 g_\alpha \sigma^{J\pi}(\alpha) P_\beta^{J\pi} - k_\beta^2 g_\beta \sigma^{J\pi}(\beta) P_\alpha^{J\pi}) = 0. \tag{21.12}$$

Actually, this equality must hold for each (J,π) term separately,since there is no particular relation between the different terms, i.e.

$$k_\alpha^2 g_\alpha \sigma^{J\pi}(\alpha) P_\beta^{J\pi} = k_\beta^2 g_\beta \sigma^{J\pi}(\beta) P_\alpha^{J\pi}. \tag{21.13}$$

Since this relation must also hold for any pair of channels α and β it follows that we can write

$$P_\beta^{J\pi} = A k_\beta^2 g_\beta \sigma^{J\pi}(\beta) \tag{21.14}$$

† The precise quantity to which this equality applies is the *collision matrix*, which is the generalisation of the collision function U_l appearing in the entrance channel formalism of Appendix F.

where the constant of proportionality A must be the same for all channels. This constant can easily be determined from (21.9), in which case (21.14) becomes

$$P_{\beta}^{J\pi}=\frac{k_{\beta}^{2}g_{\beta}\sigma^{J\pi}(\beta)}{\sum_{\beta'}k_{\beta'}^{2}g_{\beta'}\sigma^{J\pi}(\beta')} \tag{21.15}$$

where the sum in the denominator goes over all possible channels.

One important feature of equations (21.14) and (21.15) is the fact that the probability of a given *decay* mode of the compound nucleus is proportional to the cross section for *formation* of the compound nucleus out of that mode. For example, since charged-particle *capture* is hindered by the Coulomb barrier, so will *emission* of charged particles from the compound nucleus.

Combining, finally, (21.8a) and (21.15) gives for the reaction cross section

$$\sigma_{\alpha\beta}=k_{\beta}^{2}g_{\beta}\sum_{J,\pi}\left(\frac{\sigma^{J\pi}(\alpha)\sigma^{J\pi}(\beta)}{\sum_{\beta'}k_{\beta'}^{2}g_{\beta'}\sigma^{J\pi}(\beta')}\right). \tag{21.16}$$

We shall now examine the implications of factorisation and detailed balance for the two regions where the independence hypothesis is believed to be satisfied, i.e. the low-energy region of isolated resonances and the high-energy statistical region.

Breit–Wigner formula for isolated resonances. Suppose that a compound nucleus is formed out of the channel α at time $t=0$ in a definite resonant state whose peak lies at energy E_0. Since this state decays with a lifetime $\tau=\hbar/\Gamma$, the evolution in time of the wavefunction of the compound nucleus must follow the relation

$$|\psi(t)|^2=|\psi(0)|^2\exp\left(-\frac{\Gamma}{\hbar}t\right) \tag{21.17}$$

$|\psi(t)|^2$ representing the probability that the compound nucleus still exists at time t. Now the energy $\mathcal{E}$ of this state is always given by the time-dependent Schrödinger equation as

$$i\hbar\frac{\partial\psi}{\partial t}=\mathcal{E}\psi \tag{21.18}$$

so that

$$\psi(t)=\psi(0)\exp\left(-\frac{i}{\hbar}\mathcal{E}t\right). \tag{21.19}$$

It is clear now from (21.17) that $\mathcal{E}$ cannot be the real energy E_0 but rather must be regarded as complex

$$\mathcal{E} = E_0 - \frac{i}{2}\Gamma \tag{21.20}$$

so that we shall have for the wavefunction

$$\psi(t) = \begin{cases} 0 & t<0 \\ \psi(0)\exp\left[-\frac{1}{\hbar}\left(iE_0 + \frac{1}{2}\Gamma\right)t\right] & t>0\,. \end{cases} \tag{21.21}$$

These complex energies do not have any direct physical meaning, and have been introduced simply in order to accommodate the decaying states within the framework of conventional quantum mechanics†. However, we can make a Fourier expansion of the wavefunction (21.21) in terms of states having real energy:

$$\psi(t) = \int_{-\infty}^{\infty} C(E)\exp\left(-\frac{i}{\hbar}Et\right)\,dE \tag{21.22}$$

where the amplitude of the state having real energy E is

$$\begin{aligned} C(E) &= \frac{1}{2\pi\hbar}\int_{-\infty}^{\infty}\psi(t)\exp\left(\frac{i}{\hbar}Et\right)\,dt \\ &= \frac{\psi(0)}{2\pi\hbar}\int_0^{\infty}\exp\left[\frac{t}{\hbar}\left(i(E-E_0)-\frac{\Gamma}{2}\right)\right]\,dt \\ &= \frac{\psi(0)}{2\pi}\left(i(E_0-E)+\frac{\Gamma}{2}\right)^{-1}. \end{aligned} \tag{21.23}$$

Thus the probability that the compound nucleus decays into a state of real energy E is given by

$$|C(E)|^2 = \frac{|\psi(0)|^2}{4\pi^2}\left((E-E_0)^2 + \frac{\Gamma^2}{4}\right)^{-1}. \tag{21.24}$$

We can now use the detailed balance property (21.14) to calculate the cross section of *formation* of the compound nucleus out of the channel α when the bombarding energy is E: we find

$$\sigma(\alpha) \propto \frac{1}{E}\left((E-E_0)^2 + \frac{\Gamma^2}{4}\right)^{-1}. \tag{21.25}$$

† This is somewhat analogous to the use of the complex optical potential to represent absorption of entrance channel flux. The only completely correct way of representing either the compound-nucleus decay or absorption in nuclear reactions is to take account of all nucleons explicitly, and these never vanish, of course.

But since the compound nucleus must decay with a definite energy corresponding to the bombarding energy, equation (21.25) gives us the energy dependence of the entire cross section of the reaction: there is no extra energy-dependent factor associated with the decay of the compound nucleus.

The expression (21.25) for the variation of reaction cross section with energy in the vicinity of an isolated resonance is known as the *Breit–Wigner formula*, and can be rigorously proven by a much more sophisticated development than the one we have given above. It clearly has at least the right qualitative form: a peak centred at E_0 with width determined by the parameter Γ.

The peaks of the reaction cross section shown in figure 21.2, for example, are well described by the Breit–Wigner formula. On the other hand, the peaks of the *total* cross sections seen in figure 21.1, for example, show a strong asymmetry that cannot be accounted for by the presence of the $1/E$ factor in equation (21.25). This skewness arises from the fact that a significant contribution to the total cross section comes from elastic scattering and since a part of this proceeds through the optical potential rather than through compound-nucleus formation, it will not be described entirely by equation (21.25).

Hauser–Feshbach treatment of the statistical region. In this energy region so many different modes of disintegration are available to the compound nucleus that the probability of disintegration back into the entrance channel is very small, so that compound-elastic scattering can be neglected. Then with the overall direct interaction contribution also being negligible we see that the cross sections $\sigma^{J\pi}(\alpha)$, etc, for compound nucleus formation appearing in equation (21.16) can be well approximated by the corresponding total reaction cross sections, σ_{react}. Because of the denominator in equation (21.16) we shall require these total reaction cross sections for all possible entrance channels, many of which will be experimentally inaccessible†. However, because all cross sections vary smoothly in this energy region the various σ_{react} can be calculated from the optical model, provided reasonable choices are made for the optical parameters in the inaccessible channels. Thus equation (21.16), which is known as the *Hauser–Feshbach equation* in this context, enables us to calculate the cross section of a particular reaction mode from the optical model.

† For example, we have to include entrance channels in which one or both nuclei are in excited states.

PROBLEMS FOR CHAPTER IV

IV.1 Use the mass tables to calculate the Q values of the following reactions

$$^{14}N(\alpha, p)^{17}O$$
$$^{14}N(\alpha, n)^{17}F$$
$$^{11}B(\alpha, p)^{14}C$$
$$^{11}B(\alpha, n)^{14}N .$$

In the case of the endothermic reactions calculate the threshold bombarding energy.

IV.2 Express the threshold bombarding energy of a photonuclear reaction (endothermic) in terms of the Q value, and show that the result is compatible with equation (19.18), within the limits of the approximations of Appendix A.

IV.3 Consider the reaction $^{14}N(\alpha,p)^{17}O$, with a bombarding energy of 9 MeV. Calculate the energy of the protons when they emerge at an angle of (*a*) 30° (*b*) 60° with respect to the beam direction, as measured in the laboratory system. In each case calculate the energy and direction of the recoiling ^{17}O nucleus (see Appendix A).

IV.4 Verify equation (20.19) for the neutron transmission coefficient.

IV.5 The figure below (adapted from Christensen P R, Switkowski Z E and Dayras R A 1977 *Nucl. Phys.* A **280** 189) gives σ_{react} as a function of energy for the bombardment of ^{16}O by ^{12}C. Calculate the S factor for $E_{CM} = 5$ and 10 MeV.

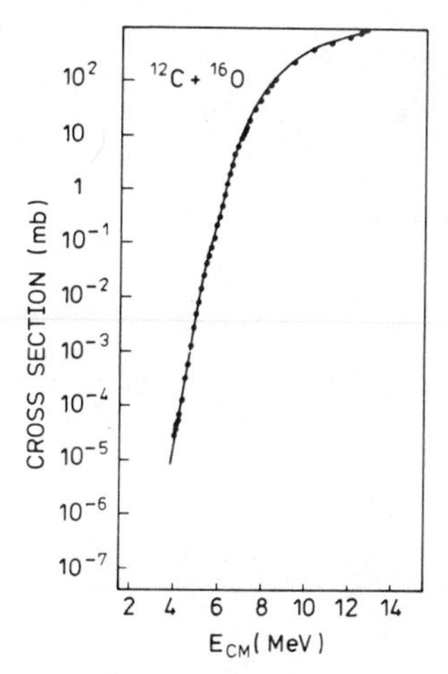

IV.6 The total cross section σ_{tot} for 3 MeV neutrons in natural cadmium is 4.5 barns. When a beam of 3 MeV neutrons is incident on a target of natural cadmium, 0.5 mg cm^{-2} thick, what fraction of neutrons will be removed from the beam? (The atomic weight of natural cadmium is 112.)

IV.7 Consider the complex square well of equation (20.25). Referring to Appendix F, show that the complex phaseshift δ for the $l=0$ partial wave is given by

$$\frac{\tan \delta}{k} = \frac{\tan \varkappa a - \varkappa \tan ka/k}{\varkappa + k \tan ka \tan \varkappa a}$$

Show then that the $1/v$ law of equation (20.2) for σ_{react} will be satisfied as $k \to 0$.

CHAPTER V

THE FISSION CHAIN REACTION

§22 NEUTRON-INDUCED FISSION

We have discussed in Chapter III the phenomenon of spontaneous fission, in which certain heavy nuclei can break up spontaneously into two comparably sized fragments, with an enormous energy release of the order of 200 MeV per nucleus. However, this process is of little interest from the point of view of a convenient source of large-scale energy, since the rate at which it occurs is far too slow in any naturally occurring nuclide. (There are, of course, artificial nuclides in which spontaneous fission is extremely rapid, but then precisely for this reason they will not be found in nature†.)

It will be remembered that the reason why spontaneous fission occurs at a finite rate rather than taking place instantaneously, as soon as the nucleus is formed, is that the changing shape of the fissioning nucleus has associated with it a potential barrier, depicted in figure 18.3, and the fission only goes through to completion by virtue of a quantum-mechanical tunnelling of this barrier. We may expect then that the rate of fission of a nucleus could be increased significantly over its spontaneous rate if it took place under the bombardment of particles bringing in a quantity of energy greater than the barrier height E_b, as shown in figure 18.3.

This has been confirmed experimentally, it being found that fission can be *induced*, as we say, by a large variety of projectiles: neutrons, protons, heavier charged particles, electrons and photons. The actual rate of fission in these circumstances will depend on the nature and energy of the bombarding projectiles, and, of course, on their flux. It should be noted that

† We recall, in fact, that it is spontaneous fission that sets an upper limit on the mass number of nuclei that can exist.

induced fission is essentially a reaction process, rather than a decay, as in the case of spontaneous fission, so that the relevant parameter is a cross section, rather than a lifetime.

Of particular interest is the case where the bombarding particle is absorbed to form a compound nucleus; this is especially likely to happen for neutrons. In this situation the excitation energy E^* of the compound nucleus will be given by

$$E^* = E_n + S_n \tag{22.1}$$

where E_n is the bombarding energy of the neutron and S_n the separation energy of a neutron from the compound nucleus, which for heavy nuclei is of the order of 5–7 MeV (see Problem III.13). It must be realised that in this case the relevant barrier height will be that of the compound nucleus and not that of the original target nucleus.

The excess of the barrier height (of the compound nucleus) over the neutron separation energy (for the compound nucleus), $E_b - S_n$, represents the necessary neutron bombarding energy E_n for fission to occur at an appreciable rate, i.e. for the barrier to be surmounted. It is found that for the isotopes ^{238}U and ^{232}Th, which constitute 99.3% and 100% of the respective naturally occurring elements, this fission threshold energy for neutrons is around 1 MeV. On the other hand, for ^{235}U, which constitutes about 0.7% of natural uranium, and the artificial nuclides ^{233}U, ^{239}Pu and ^{241}Pu, fission occurs readily even for neutrons of zero energy, because $S_n > E_b$; such nuclei are said to be *fissile*†.

The fact that the second group of nuclei are fissile and the first not is primarily a consequence of the pairing term in the semi-empirical mass formula (14.31). We notice that the compound nuclei formed from the first group will both have an odd number of neutrons, while those formed from the second group will all have an even number. Thus S_n will tend to be systematically larger for the second group, and the excitation energy E^* likewise, according to equation (22.1).

Actually, when we say that zero-energy neutrons induce fission in these nuclides, it should be realised that the same energy considerations hold when the projectiles are protons. However, Coulomb repulsion will render the fission cross section negligibly small for low-energy charged particles, and neutron-induced fission is the only kind of fission offering any possibility of large-scale energy generation‡.

† All six of these nuclei are α radioactive. However, it will be seen from table 18.1 that ^{235}U, ^{238}U and ^{232}Th are sufficiently long-lived to have survived in the earth (the rate of decay through spontaneous fission is negligible). On the other hand ^{233}U, ^{239}Pu and ^{241}Pu decay too rapidly, and lack the sufficiently long-lived progenitors to be naturally occurring.

‡ The fission cross section is much larger for higher energy protons. However, such protons are quickly slowed down in any substantial block of material to an energy where they can no longer cause fission.

In fact, the resonance-averaged cross section of fissile nuclei for neutron-induced fission obeys the famous $1/v$ law (§20), becoming very large at very low energies. However, under normal circumstances, such as those pertaining in a reactor, a lower limit on the energy of neutrons is imposed by the thermal motion. Neutrons in thermal equilibrium with their environment are said to be *thermal*, and the corresponding fission cross sections for the fissile nuclei at 20°C are shown in the first column of table 22.1. (These thermal cross sections represent, of course, an average across the Maxwell–Boltzmann distribution.)

Table 22.1 Properties of fissile nuclei for themal neutrons (20°C, averaged across the Maxwell–Boltzmann spectrum).

	σ_f, fission cross section (barns)	ν, mean number of neutrons emitted per fission	η, mean number of neutrons emitted per neutron absorbed	β, fraction of delayed neutrons (%)
^{233}U	528	2.48	2.27	0.27
^{235}U	569	2.42	2.06	0.65
^{239}Pu	785	2.86	2.06	0.22
^{241}Pu	1060	2.92	2.17	0.54

To close this section with a brief historical note, we should point out that induced fission was discovered before spontaneous fission, despite the order in which we have presented these topics in this book. Beginning with Fermi in 1934, several groups had subjected uranium to bombardment by slow neutrons, and had attempted to identify by radiochemical means the various transuranic nuclides that were expected to be produced. Such nuclides were indeed found, but there were a number of puzzling features in the scheme of observed lifetimes. Evidence accumulated that these were associated with the formation of elements considerably lighter than uranium, but since the fission process was completely unsuspected the interpretation of the experiments was slow in coming†. However, by late 1938 Hahn and Strassmann had shown conclusively that barium was formed in the bombardment of uranium, and the fission process was immediately proposed by Meitner and Frisch. Spontaneous fission was discovered in 1941.

† In fact, a German chemist, Ida Noddack, actually did propose fission as an explanation of Fermi's 1934 experiments as soon as they were announced, but her ideas were dismissed and forgotten; although Hahn had been informed of them there is no reference to Frau Noddack in his paper with Strassmann presenting the discovery of fission. Had she been taken more seriously and fission identified in 1934 rather than in 1939 there would have been the most profound military and political consequences.

§23 CHAIN REACTION

We recall now that the ratio N/Z of the β stable isobar of given mass number A increases with A (figure 14.1). It thus follows that the two fragments of the fission of any β stable nucleus will be highly neutron rich, i.e. lie high up on the excess-neutron side of the valley of stability, so that not only will the fission fragments be highly β unstable† (always β^-), but there is also the possibility that neutrons will be emitted.

That this is indeed the case had been confirmed experimentally by early 1939. Of enormous significance was the fact that there appeared to be on average more than one neutron emitted at each fission. Insofar as each of these emitted neutrons is capable of inducing fission in another fissile nucleus, it was at once realised by several groups of workers (notably von Halban, Joliot and Kowarski in France, and Fermi and Szilard in the USA) that one had here the possibility of a self-sustaining chain reaction which, beginning with a single fission event, would grow exponentially until the entire sample had been consumed. Thus the prospects of being able to obtain a large-scale release of nuclear energy were suddenly and dramatically increased: there had been much speculation on the matter ever since the discovery of radioactivity at the turn of the century, but the general opinion had been rather sceptical‡.

The fission process seemed, in fact, to be ideal from this point of view, since it combined the feature of a macroscopic chain reaction with a very high specific energy release (of the order of 200 MeV per fission). However, it was not until the following year, 1940, that sufficient data had been gathered to confirm that a chain reaction was indeed possible, and that it could both take place explosively, or be controlled, according to the circumstances. By that time military imperatives had imposed total secrecy on all fission research.

Returning to the standpoint of present knowledge, we see from the second column of table 22.1 that the mean number ν of neutrons emitted per fission from the different fissile nuclei is indeed greater than unity. (Remember that this table refers to fission induced by thermal neutrons; ν increases slightly with bombarding energy.) Most of the emitted neutrons have an energy of between 1 and 2 MeV.

The bulk of the fission neutrons are emitted within 10^{-14} seconds of fission having taken place, and are said to be *prompt*. However, a small fraction β can be delayed up to as much as a minute, a point of crucial importance for the stability of reactors (see §24). These *delayed neutrons*

† It is this that accounts for the intense radioactivity of fission waste products.
‡ With the exception of Szilard, who had conceived the idea of a neutron chain reaction as early as 1934, before the discovery of fission.

are emitted not by the primary fission fragments but by their daughter products, formed after one or more β decays† (see §17). The values of this fraction β for the various fissile nuclei are shown in the last column of table 22.1.

Actually, there is a finite probability that the compound nucleus formed by absorption of a neutron will not undergo fission but will simply de-excite to its ground state through emission of γ radiation, i.e. the neutron will be captured radiatively. Then denoting by α the ratio of neutron radiative captures to neutron-induced fissions

$$\alpha = \frac{\sigma_\gamma}{\sigma_f} \tag{23.1}$$

we see that the number of neutrons emitted per neutron absorbed will not be ν but rather

$$\eta = \frac{\nu}{1+\alpha}. \tag{23.2}$$

Now it is easy to see that if all the emitted neutrons are eventually absorbed by other fissile nuclei then each fission event will lead to η new fissions, in which case a chain reaction will indeed be possible if $\eta > 1$. That this is in fact the case for all the fissile nuclei is seen from the third column of table 22.1, where we show the measured values of η for thermal neutrons (20°C).

Of course, for the chain reaction to start at all it is necessary that there be at least one neutron already present in the sample. However, there is always a background flux of stray neutrons, coming either from cosmic rays or spontaneous fission‡.

In addition to assuming that all neutrons are eventually absorbed by fissile nuclei let us make the further assumption that they are all absorbed after the same time l. Then, beginning with a single neutron, we shall have after time l a second generation of η neutrons, after time $2l$ a third generation of η^2 neutrons, and in general, after time ml, where m is an integer, the number of neutrons present will be

$$n(t = ml) = \eta^m = \eta^{t/l} \tag{23.3}$$

assuming that the emission is always prompt.

† This means that the stability of reactors, like the stability of stars (see §27) is due essentially to the extreme weakness of the β decay interaction. (I am grateful to A Laverne for this remark.)

‡ Spontaneous fission can actually be a problem in the plutonium nuclear bomb, giving rise to a tendency to pre-ignition. This difficulty could occur if significant amounts of ^{240}Pu, which has a high spontaneous fission rate, are admixed with the predominant ^{239}Pu.

Actually, this picture is grossly over-simplified, because in reality the neutrons are not all absorbed the same time after emission. To analyse the chain reaction more carefully, let the probability that a free neutron be removed from the reaction system, i.e. reactor or bomb, in time $\mathrm{d}t$ be $\mathrm{d}t/l$: by removal we mean absorption in a nucleus (fissile or otherwise) or escape (we shall see below that the loss of neutrons through β decay is quite negligible). Then referring to §10 we see that l represents the mean lifetime of a neutron in the reaction system, i.e. it is the *mean* time that elapses between the emission of a neutron by a fissioning nucleus and its subsequent removal in one way or another.

Assuming for the moment that all neutrons are eventually absorbed by fissile nuclei and are not otherwise lost to the system, it follows that the number of fissile nuclei that absorb a neutron in the time interval between t and $t + \mathrm{d}t$ will be

$$\mathrm{d}N = n(t)\,\frac{\mathrm{d}t}{l} \tag{23.4}$$

where $n(t)$ denotes the number of neutrons present at time t. Then since for each neutron absorbed there is a net increase of $\eta - 1$ neutrons, it follows that the increase in the number of neutrons in this same time interval will be

$$\mathrm{d}n = (\eta - 1)n(t)\,\frac{\mathrm{d}t}{l} \tag{23.5}$$

provided again all the emitted neutrons are prompt. Solving equation (23.5) we have for the time variation of the neutron population

$$n = n_0 \exp\left(\frac{\eta - 1}{l}\right)t \tag{23.6}$$

which corresponds to exponential *growth* if $\eta > 1$. The fission rate, and with it the rate of energy generation, will likewise grow exponentially.

Since typical values of the lifetime l vary from 10^{-3} s in the case of a thermal reactor (§24) to 10^{-8} s in the case of a bomb (this shows how the β decay of the neutron, with its half-life of 11 min, has a negligible role to play in neutron removal), we see that the chain reaction offers the possibility of an incomparably more rapid release of the enormous amount of energy available in fission than is possible with spontaneous fission. In fact, the most obvious outcome of a chain reaction is a catastrophic explosion, as in a bomb, but we shall see how in the reactor it is possible to control the exponential growth of the chain reaction, so that we shall have a practical source of commercial energy. The relationship between the bomb and the reactor is that between a conventional explosive and an ordinary fire.

The result of our more sophisticated analysis, equation (23.6), is different from that of our preliminary crude analysis, equation (23.3). Nevertheless, the two become identical in the limit where $\eta - 1$ is very small, and in fact the possibility of regarding the neutron mean lifetime l as the interval between successive neutron generations will be very useful when we discuss the question of reactor control later in this section. Indeed, l is often referred to as the *generation time*.

It might be thought that ^{238}U, the dominant isotope in natural uranium, could also support a chain reaction, since the energy of many of the neutrons emitted in fission exceeds the threshold energy of 1 MeV. However, mainly because of inelastic collisions with uranium nuclei, the neutron's energy rapidly falls below threshold. Thus, only fissile nuclei can sustain a chain reaction†, so that in natural uranium one depends entirely on the ^{235}U, which is thus the only naturally occurring nuclide in which a chain reaction is possible.

Neutron losses. We have remarked that the number η of neutrons emitted per neutron absorbed in a fissile nucleus is considerably larger than unity, so that a chain reaction is in principle possible, since each fission event will lead to η new fissions. Actually, as we shall discuss below, losses of various kinds occur, so that not every emitted neutron will be absorbed by another fissile nucleus. Thus for each neutron that is absorbed in a fissile nucleus the mean number of neutrons that are emitted and go on to be absorbed in other fissile nuclei is not η but rather some number k, where $k < \eta$.

Since each fission will now lead to k new fissions the critical condition for a chain reaction will be

$$k > 1. \qquad (23.7)$$

Also with k replacing η in equation (23.6) we have for the time variation of the neutron flux

$$n = n_0 \exp\left[\left(\frac{k-1}{l}\right)t\right]. \qquad (23.8)$$

We call k the *effective multiplication factor*.

Without this reduction of the effective value of η, it would be impossible to make any practical application of the chain reaction. The triggering of a bomb, for instance, depends on suddenly increasing k from below the *critical value* of unity to above, while the control of a reactor involves holding k at exactly unity. The different ways in which the value of k can actually be controlled will become apparent in the following discussion of the different sources of neutron loss.

One of the most significant sources of neutron loss is through the surface of the assembly. The relative importance of this will diminish as the surface/mass ratio decreases, i.e. as the overall size of the assembly increases for a given density, or as the size *decreases* for a given mass. For

† See, nevertheless, our discussion of the fission–fusion–fission bomb (§30).

this reason a chain reaction cannot take place unless the assembly is larger than a certain critical size, or unless its density exceeds a certain critical value. Thus the so-called gun-type of bomb consists of two (or more) pieces of fissile material, each of subcritical size, brought together very rapidly by a chemical explosion to form one piece of supercritical size, thereby bringing about the required sudden increase in k. The Hiroshima bomb, which used ^{235}U as the fissile nuclide, was of this kind.

A quite distinct type of fission bomb uses a single piece of fissile material in the form of a sphere surrounded by chemical explosives. Criticality is then achieved by implosion, the fissile charge being momentarily compressed to maybe three quarters of its normal volume. The Nagasaki bomb, which used ^{239}Pu, was of this kind.

We see that for the first type of bomb, where the density remains constant, there is a definite critical mass and hence a lower limit on the explosive yield. However, for the implosion type the critical mass will depend only on the densities that can be achieved, so that in principle there will be no lower limit on the explosive yield of this type.

On the other hand, there is quite clearly an *upper* limit on the explosive yield possible with the implosion bomb, since the mass of the charge must be lower than the critical mass for normal densities. As for the gun-type of fission bomb, it would appear that its yield could be increased indefinitely by increasing the number of pieces of fissile material, but the engineering difficulties of synchronising their rapid assembly means that for this type of bomb also there is a practical upper limit on the yield. This was one of the key factors lying behind the decision to develop the hydrogen bomb† (§30), which suffers from no such limitations.

It is also because of surface losses that a reactor has to exceed a certain size before it can sustain a chain reaction‡. This has implications for what would happen in the event of too rapid an exponential growth of the chain reaction: as the energy flux grew the reactor would simply melt and fall apart into small pieces which would be of subcritical size, whereupon the chain reaction would stop. Thus there can be no question of a reactor blowing up as a full-scale nuclear bomb, although there is always the possibility that the heat release would be large enough to breach the containment vessel, with a catastrophic release of radioactivity into the atmosphere or ground.

A special kind of surface loss arises in reactors constructed primarily for research purposes: these reactors are usually provided with beam-holes which will act as a uniquely intense source of neutrons.

† Known as the H bomb. Most confusingly, the term 'atomic bomb' (A bomb) is confined in popular usage to the fission bomb.

‡ The critical size of a reactor can be reduced by surrounding it with a material that reflects neutrons.

The other main source of neutron loss is through absorption by non-fissile nuclei that may be present (it will be recalled that the loss through β decay is negligible). A prime example of this occurs in a reactor using natural uranium as fuel. We have already remarked that the predominant isotope, ^{238}U, is non-fissile and cannot sustain a chain reaction. It turns out, however, that the ^{238}U is not simply passive, but actually plays a detrimental role, since between 6 and 200 eV, below the fission threshold, the cross section for radiative capture of neutrons shows a series of very strong resonances (figure 21.1). Thus, as fission neutrons are being slowed down by successive collisions with uranium nuclei, a large fraction is lost by non-fission absorption in ^{238}U, before reaching the thermal energies for which the cross section for fission with ^{235}U is maximal, on account of the $1/v$ law. It is for this reason that a block of natural uranium, no matter what its size, can never sustain a chain reaction.

It would appear, then, that the uranium will have to be considerably enriched in the ^{235}U component, a process that is extremely expensive because of the great difficulty of separating isotopes of heavy elements. However, while some degree of enrichment is necessary for many types of reactor, and in particular for the bomb†, we shall see in §24 that it is possible to establish a chain reaction in natural uranium, provided it is not used as a single block but is rather mixed with what is called a 'moderator'.

Neutron loss through absorption in non-fissile nuclei occurs also in certain waste products of the fission process (as these accumulate in a reactor they eventually stop the chain reaction and necessitate the removal of the fuel elements long before all the fissile material has been consumed: for this reason they are known as 'poisons'); in samples that have been placed in the reactor with the express intention of being transmuted into some other isotope, either for activation analysis of the sample itself or for the production of radioisotopes; and in the various components of the reactor (or bomb).

In connection with this last point it is to be noted that neutron absorption is introduced intentionally into a reactor as a means of controlling it, i.e. of varying k. This is effected by inserting rods of cadmium or boron, both of which have a very high non-fission absorption cross section for neutrons, into the reactor. Usually these *control rods* will be in such a position that

† It should be realised that the chain reaction in the bomb is sustained mainly by so-called *fast* neutrons, i.e. neutrons whose energy is considerably greater than thermal, despite the fact that the fission cross section is greatest at thermal energies. The bomb is designed in this way since with thermal neutrons the generation time l would be too slow for a violent explosion to ensue. Thus the fissility of ^{235}U, i.e. the fact that it has zero fission threshold, is not essential to the functioning of the bomb: it is sufficient that fission still be possible after the emitted neutrons have suffered several collisions. It is because this latter condition is not satisfied in ^{238}U that a chain reaction cannot be sustained therein.

the multiplication factor k is exactly unity. If it is wished to increase the level of activity, the rods will be withdrawn so that k becomes larger than unity, in which case the fission rate will begin to rise exponentially. When the required level of activity has been reached the rods will be returned to their original configuration, corresponding to $k = 1$. The all important question of the stability of this control will now be discussed.

Reactor control: delayed neutrons. Let us consider more closely what happens when we attempt to raise the level of activity of a reactor by withdrawing the control rods, thus increasing k over unity. According to equation (23.8) the neutron density will begin to rise exponentially, thus

$$n = n_0 e^{t/T} \tag{23.9}$$

where we have introduced the *reactor period*

$$T = \frac{l}{k-1}. \tag{23.10}$$

This is the time in which the neutron flux increases by a factor of e. As a typical example, suppose that k is increased from unity to 1.01. Since l is never longer than 10^{-3} s it follows that T will be at most 0.1 s, so that after 1 s the neutron flux will have increased by a factor of over 20 000.

Clearly, there would seem to be a significant danger that the reaction rate would rise so rapidly as to reach a disastrous level before the control rods could be restored to their $k = 1$ configuration, unless the adjusting mechanism had a remarkable degree of precision and speed of response. We shall now show that because a certain fraction β of the fission neutrons are delayed, a fact that was not taken into account in our simplified theory of the chain reaction given above, there is a considerable margin of safety for error in the setting of the control rods.

The details of these delayed neutrons for the specific case of ^{235}U are shown in table 23.1, where it will be seen that there are several different groups of such neutrons, each, with its characteristic half-life, being associated with a particular nuclide in the decay chain of the fission products. Averaging over these groups we find that the mean half-life of the delayed neutrons is about 9 s, and since they represent a fraction $\beta = 0.65\%$ of the total neutron emission it follows, using equation (10.7), that the neutrons are emitted with an average delay $\bar{\delta}$ of about 0.08 s. Thus the *effective* generation time, i.e. the time that elapses between the absorption of a neutron by a fissile nucleus and the subsequent absorption of the k available neutrons of the next generation, is increased from l to $l + \bar{\delta} \simeq \bar{\delta}$. Equation (23.9) should thus be replaced by

$$n(t) = n_0 e^{t/T^*} \tag{23.11}$$

where we have introduced the *effective* reactor period

$$T^* = \frac{\bar{\delta}}{k-1} \tag{23.12}$$

in place of equation (23.10). Then with $k = 1.01$ again, for example, the period will be about 8 s, rather than the 0.1 s that we found when we assumed that all the fission neutrons were prompt. The fact that a certain fraction of the fission neutrons is delayed is thus seen to lead to a dramatic increase in the controllability of the reactor.

Table 23.1 Groups of delayed neutrons in fission of ^{235}U by thermal neutrons.

$t_{1/2}$ (s)	Yield (%)
55.7	0.0215
22.7	0.1424
6.22	0.1274
2.30	0.2568
0.61	0.0748
0.23	0.0273

Actually, this treatment of the role of the delayed neutrons is greatly over simplified, since it assumes that *all* the fission neutrons are emitted with the same delay time of $\bar{\delta} \simeq 0.08$ s. If this were indeed the case then the chain reaction could never be violently explosive and there would be no fission bomb!

To repair this obvious defect in our treatment we define effective multiplication factors for the prompt and delayed neutrons separately

$$k_d = \beta k \tag{23.13a}$$

$$k_p = (1-\beta)k \tag{23.13b}$$

so that

$$k_p + k_d = k\ . \tag{23.14}$$

If now $k > 1/(1-\beta)$ we shall have $k_p > 1$, so that the reacting system (reactor or bomb) will be super-critical to the prompt neutrons alone, and the delayed neutrons will have no stabilising effect at all. According to this improved picture, which is still highly simplified, a reactor will be super-critical, but still controllable, if

$$1 < k < \frac{1}{1-\beta}\ . \tag{23.15}$$

§24 THERMAL REACTORS: MODERATOR

We recall now that one reason why a chain reaction cannot be established in a block of natural uranium is that as the fission neutrons are being slowed down to thermal energies by virtue of their collision with uranium nuclei a very large fraction suffer non-fission absorption in the predominant ^{238}U nuclei. We recall also that this absorption is a resonant process, taking place when the neutron energy lies between 200 and 6 eV. Thus if a neutron can avoid being absorbed in ^{238}U while it slows down through this energy interval it is unlikely to suffer non-fission absorption thereafter, and instead will probably continue to lose energy and become thermalised, at which point the probability of absorption in fissile ^{235}U becomes maximal, because of the $1/v$ law.

It would appear, then, that if the fast neutrons could be prevented from colliding with ^{238}U nuclei and instead be slowed down to thermal energies by some other means, there would be some hope of establishing a chain reaction. One obvious way of doing this is to mix the uranium fuel with some other material, known as the *moderator*, the nuclei of which do not absorb neutrons strongly. This is the principle of the *thermal reactor*.

Unfortunately, while the presence of the moderator reduces the fraction of neutrons suffering non-fission absorption, it increases the bulk of the reactor and thus leads to a greater surface loss. Since it is also inevitable that there is some absorption in the moderator it is clear that the less the required amount of moderator the better. Now for a given concentration of moderator material the fraction of neutrons that will be lost through absorption by ^{238}U will be proportional to the number of collisions that a neutron makes in slowing down to thermal energies. Thus, since the fractional loss of neutrons that can be tolerated is fixed, we see that the amount of moderator required will be smaller the fewer the number of collisions that the neutron makes in becoming thermalised, i.e. the greater the energy loss at each collision.

Now an elementary calculation in classical mechanics shows that the loss of kinetic energy that a moving body suffers when it collides with a stationary body is greater the lighter the struck body. This suggests that hydrogen would be the best moderator, and indeed many modern reactors, particularly American commercial reactors, do use this, in the form of ordinary water ('light water'), which can also serve as the coolant. However, hydrogen has the disadvantage of absorbing neutrons, so if a chain reaction is to be maintained the uranium fuel must be enriched in ^{235}U to at least 1%.

Alternatively, one could consider other materials as moderators. After hydrogen, the most likely candidate is deuterium, and it is found that this has a much lower cross section for neutron absorption, so that a chain reaction can be set up in natural uranium without any enrichment being necessary. This has served as the basis for many successful reactors, notably the Canadian CANDU, the deuterium moderator being in the form of heavy water. Of course, the production of heavy water involves an expensive separation of isotopes, but this is easier than the separation of heavy isotopes involved in uranium enrichment.

The only other substance besides hydrogen and deuterium that has been used extensively as a moderator is carbon, in the form of graphite, a material that is abundantly available and requires no isotope separation. The world's first *artificial* reactor, built by Fermi in Chicago in 1942, had a graphite moderator, as do the British gas-cooled reactors. In fact it had previously been shown that a *homogeneous* mixture of natural uranium and graphite could never support a chain reaction, no matter what were the proportions or size of the system: the maximum possible value of k in these circumstances was found to be 0.8. However, Fermi and Szilard showed that k could be increased significantly in a *heterogeneous* system, consisting of lumps of uranium embedded in a graphite lattice. Fermi's reactor consisted simply of a pile of graphite bricks, some of which contained uranium, and thus became known as a 'pile'.

A natural, prehistoric reactor. For many years it was believed that the fission chain reaction was essentially a man-made phenomenon, and that therefore Fermi's 1942 reactor represented the site of the first chain reaction ever to take place, at least on earth (the bomb came in 1945). However, in 1972 it was found that uranium ore from the Oklo mine in Gabon was significantly deficient in ^{235}U. Hitherto the concentration of ^{235}U in all known sources of natural uranium, both terrestrial and extraterrestrial, had been found to be remarkably constant at 0.720%, but the samples from Oklo showed considerable variations, some being as low as 0.4%.

A detailed study showed that the missing ^{235}U had simply been consumed in a fission chain reaction, the conditions at some time in the past having been such that this could take place naturally. To be more precise, the Oklo deposits were laid down some 1700 million years ago, at which time the concentration of ^{235}U in natural uranium was about 3%, the present value being much lower because ^{235}U α decays much more rapidly than ^{238}U (see table 18.1). With this concentration a chain reaction would be possible with ordinary water, present in the form of groundwater, serving as moderator. Of course, certain geometric conditions relating to the size, shape and configuration of the deposit would have to be satisfied, and it is for this reason, in fact, that the phenomenon is not more widespread†.

† For more details see A natural fission reactor, G A Cowan, *Scientific American* **235**(1) (July 1976) p. 36.

§25 CONVERSION AND BREEDING

We have pointed out that of the four fissile nuclides only one, ^{235}U, is found in nature. The other three, ^{233}U, ^{239}Pu and ^{241}Pu, can be obtained only artificially, by means of nuclear reactions. However, if these nuclides were to serve as a viable reactor fuel they would have to be produced in quantities far in excess of what is possible in the usual laboratory nuclear reaction experiment.

Nevertheless, there is one situation in which nuclear reactions can occur on a sufficiently large scale and that is in a nuclear reactor itself, with its enormous neutron flux. The problem then is to find nuclides that can be transmuted into fissile nuclides by neutrons, with the condition that they occur naturally in fairly large quantities, if the process is to be of any practical interest. It turns out that both ^{239}Pu and ^{233}U can be produced in this way, ^{238}U being used for the former and ^{232}Th for the latter. In both cases the relevant neutron reaction is a capture, followed by two β decays†, thus

$$^{238}\text{U} + \text{n} \rightarrow {}^{239}\text{U} \xrightarrow[23.5\ \text{min}]{\beta^-} {}^{239}\text{Np} \xrightarrow[2.3\ \text{d}]{\beta^-} {}^{239}\text{Pu} \tag{25.1}$$

$$^{232}\text{Th} + \text{n} \rightarrow {}^{233}\text{Th} \xrightarrow[22.2\ \text{min}]{\beta^-} {}^{233}\text{Pa} \xrightarrow[27.4\ \text{d}]{\beta^-} {}^{233}\text{U}\,. \tag{25.2}$$

Thus, although the nuclides ^{238}U and ^{232}Th are not themselves fissile, they can be converted into nuclides that are fissile by being suitably placed in the neutron flux of a reactor. These nuclides are said to be *fertile*, and the process of transforming them to fissile nuclides is known as *conversion* (or sometimes loosely as 'breeding', although we shall follow the usual practice of using this in a more specialised sense, as will be discussed below).

As for the remaining fissile nuclei, ^{241}Pu will be formed along with ^{239}Pu if the neutron bombardment of the ^{238}U is sufficiently prolonged: two successive captures of neutrons by the ^{239}Pu lead first to the formation of ^{240}Pu and then ^{241}Pu; note, however, that ^{241}Pu is β unstable and is of no particular interest as a reactor fuel. On the other hand, the one fissile nuclide that does occur naturally, ^{235}U, cannot be produced artificially in large quantities, i.e. by neutron bombardment, from naturally occurring nuclides.

Production of ^{239}Pu takes place automatically in all uranium-fuelled thermal reactors, the parasitic absorption of neutrons by ^{238}U that one tries to minimise as much as possible by means of the moderator being precisely the reaction that generates ^{239}Pu. Some of this ^{239}Pu undergoes fission in

† The times shown in these disintegration schemes are half-lives.

the reactor, and adds to the total power output over and above the power generated by the ^{235}U. However, significant quantities of ^{239}Pu remain in the used fuel elements after normal operation, and can be extracted by chemical methods. But because of the many hazards associated with it, such *reprocessing*, as it is called, is far from being accepted as a matter of routine procedure, and most of the world's stock of ^{239}Pu remains unextracted in the used fuel rods and wastes that are accumulating.

At the present time, the relatively small amount of ^{239}Pu that has been extracted is being used mainly for weapons†. The long-term prospect is that ^{239}Pu will be extracted to serve mainly as a fuel for the fast-breeder reactor (see below; remember that ^{235}U cannot itself be produced on a large scale). In principle, the extracted ^{239}Pu can also be used in thermal reactors, and in particular can serve to enrich the fuel of those thermal reactors that cannot burn natural uranium, e.g. light-water reactors. In this way, one can eliminate the expensive process of isotope separation that is involved in ^{235}U enrichment.

As for ^{232}Th, its conversion to ^{233}U can be effected simply by placing appropriate amounts in an ordinary uranium-fuelled reactor. So far, this interesting approach has not been developed to the point of commercial viability.

As the fissile nuclei ^{239}Pu (or ^{233}U) are formed they can be fed back into the chain reaction (either in the same reactor or another) to 'seed' a new cycle of conversion, provided there remains sufficient fertile material. Clearly, when man first began to generate the artificial nuclides ^{239}Pu and ^{233}U in the 1940s, the process had to start with reactors fuelled by ^{235}U, the only available fissile nuclide. However, as the stocks of ^{239}Pu and ^{233}U accumulate they can take over from the diminishing supply of ^{235}U‡, so that the long-term trend will be for ^{239}Pu and ^{233}U to regenerate themselves, or each other.

Breeding. Since η neutrons are emitted for each neutron absorbed in a fissile nucleus, and since just one of these neutrons has to be absorbed in another fissile nucleus for the chain reaction to be propagated, it follows that there remain $\eta - 1$ neutrons available for converting fertile nuclei. Thus if there were no other losses it would be possible to convert $\eta - 1$ fertile

† The first wartime reactors were constructed with the express purpose of producing not energy but rather ^{239}Pu for bombs. The main advantage of ^{239}Pu over ^{235}U as an explosive is the relative ease with which it can be isolated, no isotopic separation being necessary (note, nevertheless, that the chemistry of the plutonium separation is very sophisticated). It will be seen that possession of a reactor goes a long way towards enabling the owner to make a bomb.

‡ The large stocks of ^{239}Pu that would become available if there were any extensive nuclear disarmament could be used in the same way.

nuclei for each fissile nucleus consumed. In practice there will always be some loss (surface leakage and absorption by components), so defining the *conversion ratio* r as the number of fertile nuclei converted per fissile nucleus consumed, we have

$$r \leqslant \eta - 1 . \tag{25.3}$$

Now we see from table 22.1 that the value of η is always just a little over 2, so that there exists the possibility of having $r > 1$, i.e. for each fissile nucleus consumed more than one fertile nucleus would be converted. By suitable recycling it will then become possible to convert *all* fertile material into fissile material. This situation of conversion with $r > 1$ is called *breeding*. It opens up the prospect of being able to convert all the world's uranium and thorium (which is 100% ^{232}Th) into fissile fuel. Now ^{238}U is 140 times more abundant than ^{235}U, and the world's supply of thorium is estimated to be three times as great as that of uranium, so we see that breeding offers the possibility of increasing the supply of fission energy by a factor of several hundred†, since the breeding itself requires no energy expenditure at all.

On the other hand, if r is less than unity the total number of fertile nuclei that it will be possible to convert, starting with N_0 fissile nuclei, is

$$N = N_0 (r + r^2 + r^3 + \ldots)$$

$$= N_0 \frac{r}{1 - r} \tag{25.4}$$

successive terms in the first line corresponding to successive conversion cycles. (Actually, the total mass of fertile meterial, ^{238}U and ^{232}Th, that can be found in the earth is not infinite, so its complete conversion will not rigorously require $r > 1$. Nevertheless, it is several hundred times greater than the original mass of fissile material, ^{235}U, so that complete conversion will require that r be very close to unity indeed.)

Because of neutron losses the inequality is always applicable in equation (25.3), and a more realistic value that η must take in order to have breeding, i.e. $r = 1$, is $\eta \simeq 2.2$ rather than 2.0. Thus it is clear from table 22.1 which, note, refers to thermal neutrons, that breeding will be impossible with either ^{235}U or ^{239}Pu in a thermal reactor. The best value of r that is possible with a thermal reactor burning ^{235}U is around 0.7 (this is with heavy-water moderation, as in the CANDU), which means that no more than 2% of the ^{238}U is converted.

However, referring to figure 25.1, we see that η has a general tendency to increase with neutron bombardment energy, and that for both ^{235}U and

† Further breeding is obviously impossible once all fertile material has been converted into fissile material; it is sheer journalistic nonsense to suggest that the process of 'generating as much fuel as is consumed' can go on indefinitely.

^{239}Pu breeding should be possible with fast neutrons, i.e. neutrons that have suffered little slowing down after their emission during fission. Clearly, instead of adding a moderator to slow down the neutrons as rapidly as possible, we should dispense with it and rather take steps to *prevent* the neutrons from being slowed down. But we have already seen (§23) that if we do this in natural uranium then too many neutrons are captured in ^{238}U and the chain reaction will not propagate. Nevertheless, if the uranium is considerably enriched in ^{235}U, then, as we have already indicated, and as indeed happens in a bomb, a chain reaction should be possible without a moderator.

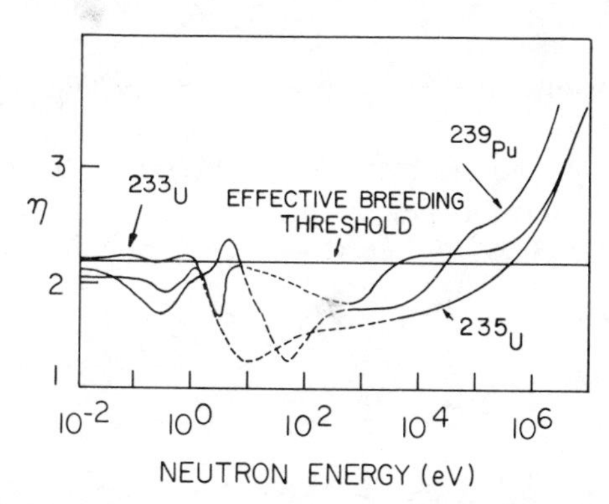

Figure 25.1 Variation of η with neutron bombarding energy. Taken from an unpublished Chalk River report (A J Mooradian, CRNL-1056).

This, then, is the key to the breeding of ^{239}Pu: we build a reactor burning enriched uranium (to at least 20%) without a moderator. Such a so-called 'fast reactor' is conceptually simpler than a thermal reactor, and indeed its principle and breeding capability were understood from the earliest times. However, there are enormous technical difficulties involved, and it is for this reason that fast breeding has made such slow progress in being implemented commercially. In the first place, these reactors are inherently less stable than thermal reactors because the fraction β of delayed neutrons is less for ^{239}Pu than for ^{235}U (see table 22.1). Also, because of the $1/v$ law, the control rods will be less effective in a fast reactor.

Another problem arises from the fact that with there being no moderator the volume per unit mass of fissile material will be much less in a fast reactor, with the result that for a given total power output the power generated per unit volume will be much greater than in a thermal reactor, so that heat extraction becomes much more difficult. At the same time, because we now have to reduce moderation to a minimum, the coolant must consist of as *heavy* an element as possible, and water is specifically excluded. The best approach appears to be with molten sodium.

By way of contrast we conclude by noting that breeding in a thermal reactor *may* be possible for ^{233}U: we see from table 22.1 that $\eta - 2$ for thermal neutrons is significantly higher with this nucleus than with either ^{239}Pu or ^{235}U. Much research remains to be done, but the prospect of being able to completely convert thorium, which, remember, is some three times more abundant than uranium, and without recourse to the technically uncertain fast breeder, is of the greatest interest to our energy future.

BIBLIOGRAPHY

Bell G I and Glasstone S 1970 *Nuclear Reactor Theory* (New York: Van Nostrand)

Inglis D R 1973 *Nuclear Energy: Its Physics and Its Social Challenge* (Reading, MA: Addison-Wesley)

Zweifel P F 1973 *Reactor Physics* (New York: McGraw Hill)

CHAPTER VI

THERMONUCLEAR FUSION

§26 GENERAL INTRODUCTION TO THERMONUCLEAR FUSION

In this last chapter we turn to a second kind of self-sustaining process, in which nuclear reactions can occur on a scale incomparably more vast than is possible even with induced fission. In discussing this latter topic in the previous chapter it was pointed out that the existence of the Coulomb barrier constituted a most serious obstacle to the realisation of charged-particle reactions on a macroscopic scale, without the use of accelerators to produce beams of projectiles, and it is certainly true that the fission chain reaction can only proceed through the intermediary of neutrons.

However, if we consider any body of gas and imagine its temperature to be increased indefinitely, then it is clear that with the increasing kinetic energy of thermal motion of the molecules† of the gas there will be an ever greater probability of penetrating the Coulomb barrier between the nuclei, until the point will be reached where nuclear reactions can take place at a significant rate throughout the body of the gas, or at least in those regions where the temperature is sufficiently high. Such *thermonuclear reactions*, as they are called, will proceed more readily between light nuclei than between heavy nuclei, because of their lower Coulomb barrier.

Now from figure 14.2 it is seen that reactions between light nuclei leading to the formation of heavier nuclei will be accompanied by a release of energy, i.e. they are exothermic. As specific examples of such *fusion*

† At temperatures where nuclear reactions take place molecules are completely dissociated into their component electrons and nuclei.

reactions we mention

$$^{2}H + {}^{2}H \rightarrow {}^{4}He + 23.84 \text{ MeV} \quad (26.1a)$$

$$^{12}C + {}^{12}C \rightarrow {}^{24}Mg + 13.93 \text{ MeV} . \quad (26.1b)$$

Thus once the temperature in some region of a gas of light nuclei has risen to the point where fusion reactions can be ignited, the process will tend to be self-sustaining: the energy release maintains the temperature at the required level and heats the cooler regions of the gas, so that the fusion process will spread throughout the body of the gas.

Thus thermonuclear fusion, like the fission chain reaction, is characterised by the release of energy on the macroscopic scale. However, there is the basic difference that while the energy release is essential to the propagation of thermonuclear fusion, it is incidental to the propagation of the fission chain reaction, this being, of course, propagated by the emitted neutrons. Also, in releasing energy the two processes exploit different branches of the binding-energy curve of figure 14.2, both approaching the minimum at ^{56}Fe, but from opposite sides.

Thermonuclear fusion is the principle of the so-called hydrogen bomb, the high temperature necessary for ignition of the reactions being produced by the detonation of a fission bomb. Now one may well wonder whether thermonuclear fusion, like the fission chain reaction, can take place not only explosively, but also quasistatically, as in a fire. In fact, at the present time a vast research effort is being devoted to the possibility of controlling thermonuclear fusion and using it as a source of commercial energy, but so far it has not proven possible to reach the required ignition temperature non-explosively.

Nature has already succeeded in harnessing thermonuclear fusion, for this is the way in which all the stars generate virtually all their energy, right up to the point where they die, either peacefully as a white dwarf, or violently as a supernova. As we discuss in Appendix D, a newly-born star, consisting mainly of hydrogen, contracts gravitationally, with the result that its temperature rises. Eventually the central part of the star will be hot enough for the nuclear reactions that convert hydrogen into helium to begin, and the rate of these reactions will increase rapidly as the temperature of the contracting star continues to rise. When the rate of nuclear energy release is equal to the rate at which the star is radiating energy out into space, the star will stabilise, i.e. both the contraction and the heating-up will cease. The star will remain in this stable condition for many millions of years, shining entirely through the conversion of hydrogen to helium in the core of the star (such stars are described as 'main-sequence'). As for the helium that is formed, the Coulomb barrier for this is four times higher than for hydrogen, so that it will not burn at the temperature where the hydrogen-burning star has stabilised. However, after a significant amount of hydrogen has been burnt the rate of nuclear energy generation will fall, whereupon gravitational contraction will resume, and continue until the attendant temperature increase leads to helium burning in the core of the star on a sufficient scale for stability to be established once more. (Actually, the helium-burning core will be surrounded by a shell of burning hydrogen.)

We see thus that the overall history of the star is one of gravitational contraction punctuated by phases of nuclear burning, in which heavier and heavier nuclei are synthesised out of the primordial hydrogen, the ashes of one phase being the fuel of the next. A limit to this process is set by the minimum at ^{56}Fe in the binding-energy curve of figure 14.2, but if and when that point is reached the star will explode as a supernova†. In the resulting violent non-equilibrium conditions all the stable nuclides that have not already been synthesised‡, including the uranium and thorium isotopes, will be formed very rapidly and dispersed throughout the galaxy along with all the nuclei that were produced over hundreds of millions of years by thermonuclear fusion.

Thus the stars are not only the ultimate source of *all* our energy (remember that they synthesise all the fissile isotopes)§ but also of almost all the chemical elements heavier than hydrogen (actually, a large amount of helium was formed in the original big bang). This enormous significance of nuclear reactions, as they occur in the stars, is the main theme of this chapter and, in a sense, of the entire book.

Our discussion of stellar evolution in this chapter will be limited to a brief qualitative description of what are believed to be the most important sequences of nuclear reactions that are followed, but in Appendix E we derive general expressions for the rate of thermonuclear reactions as a function of temperature. For a discussion of the hydrostatic aspects of stellar evolution, and in particular of the way in which at each stage of nuclear burning the star stabilises with a definite radius and central temperature, the reader is referred to Appendix D. But to have anything resembling a complete picture, the reader should consult the references given in the bibliography.

† Lighter stars may never reach the stage of forming ^{56}Fe, and rather than blow up as supernovae they will die tranquilly as white dwarfs (Appendix D).

‡ Note that long before the star reaches the supernova stage, many nuclei much heaver than ^{56}Fe will be produced by the neutron-capture s process (§29).

§ Thermonuclear reactors, if they can ever be developed, will represent the first source of energy that does not depend ultimately on the stars. But in a sense they will be stars themselves!

§27 HYDROGEN BURNING IN STARS

A star newly condensed out of the interstellar gas consists mainly of ^{1}H, together with some admixture of heavier nuclei formed either in the primordial big bang (in the case of very light nuclei such as ^{2}H and ^{4}He) or in the interior of other stars that have in one way or another expelled matter. One obstacle to the fusion of hydrogen into heavier nuclei is that all stable nuclei heavier than hydrogen must contain neutrons, and since the proton is stable against β^+ decay into neutrons there arises the question as to where the necessary neutrons can come from.

One obvious source for such neutrons lies in the small amounts of heavier nuclei that are already present in a young star. The simplest possibility here is the reaction between hydrogen and deuterium

$$^1\text{H} + {}^2\text{H} \rightarrow {}^3\text{He} + 5.49\ \text{MeV}\ . \tag{27.1}$$

This reaction does indeed occur during the initial contraction of a star, but there is too little deuterium to permit any substantial burning of hydrogen in this way†.

In fact, since all the heavier nuclides in the new star are much less abundant than hydrogen, it is clear that they can serve in the fusion of large amounts of hydrogen only insofar as they act as catalysts. That is to say, the fusion reactions must form a network in which the original heavy nuclides emerge at the end, along, of course, with the products of hydrogen fusion: this means that some β^+ decays must occur in the sequence of reactions. Such a network, the so-called CNO tri-cycle, has indeed been found, but before discussing it we first consider a process which allows hydrogen fusion to take place even in the initial absence of all heaver nuclei, despite the β^+ stability of the proton.

Proton–proton chain. Even though an isolated proton cannot undergo β^+ decay to a neutron, Bethe pointed out in 1939 that one of an interacting *pair* of protons could do so, if the bombarding energy were high enough. Thus the following reaction fusing together two hydrogen nuclei becomes possible, even if no heavier nuclei are present:

$$^1\text{H} + {}^1\text{H} \rightarrow {}^2\text{H} + e^+ + \nu + 0.42\ \text{MeV} \tag{27.2}$$

(the formation of free neutrons rather than deuterium will have negligible probability at the temperatures typical of hydrogen-burning stars).

† Nevertheless, this reaction is significant in that it leads to deuterium having a lower concentration in the interior of a star then in the interstellar medium, a situation which implies that deuterium must have been formed in the big bang, and that stars actually destroy it rather than create it.

This reaction is extremely slow, since in addition to the usual Coulomb barrier, common to all charged-particle reactions, there is also the weakness of the β decay process to contend with. The cross section of this reaction is, in fact, so small that there is no question of its ever being measured directly, but it can be calculated quite reliably, since we have a fairly complete knowledge of all the different factors involved, both concerning the nuclear wavefunctions and the details of the weak interaction process. The value obtained for the cross section is about 10^{-27} barn at 5 keV, the effective thermal energy for hydrogen-burning stars with a typical central temperature of around 10^7 K (see Appendix E).

With a steady, if somewhat weak, supply of deuterium being assured, the reaction (27.1) can now proceed indefinitely, until no more hydrogen is left. The ^{3}He that is formed in this way can react in three different ways to form ^{4}He, thus

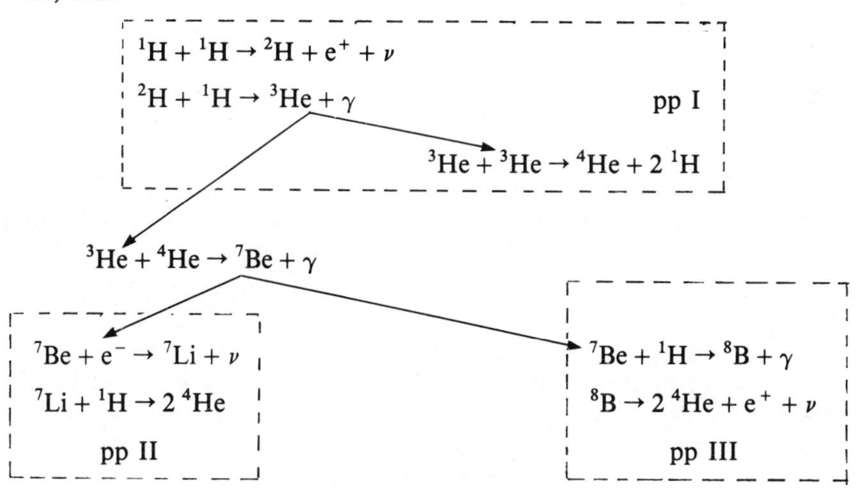

The process that we have labelled pp I is the only one that is possible in a star consisting of pure hydrogen. However, after the build-up of ^{4}He, or right from the beginning if ^{4}He was already present, we can have formation of ^{7}Be, and the pp II and pp III processes become possible.

In all three of these processes we see that formation of one ^{4}He nucleus involves two β transitions, since in each case we have in effect the reaction†

$$4\,^1\mathrm{H} \rightarrow {}^4\mathrm{He} + 2e^+ + 2\nu + 24.69\ \mathrm{MeV}. \qquad (27.3)$$

† This energy release of 24.69 MeV will, in fact, be augmented by the 2.04 MeV coming from the annihilation of the two β positrons on the star's electrons. But then it must be reduced by the average energy of the two neutrinos, since these, unlike the annihilation photons, usually pass right through the star without depositing their energy; this average neutrino energy will be different in each of the three different pp chains.

In pp I there are two reactions of type (27.2), in pp II one such reaction and one electron capture, while in pp III there is again one such reaction and one β^+ decay. It is to be noted that the neutrinos coming from the reaction (27.2) are not sufficiently energetic to be detected by Davis' ^{37}Cl experiment (see §11 and Problem III.8).

The Coulomb barrier for the different reactions in which ^3He is transformed into ^4He, either directly or indirectly, is higher than the Coulomb barrier involved in the *formation* of ^3He, so that it might be thought that the hydrogen-burning star would stabilise with the formation of ^3He rather than of ^4He. However, this is not possible since relatively little energy is released in the formation of ^3He and the star continues to contract and heat up until ^3He burning begins on a significant scale. (Note that the energy release per nucleon in reaction (27.3) is some five times greater than in uranium fission.)

CNO tri-cycle. We pointed out at the beginning of this section that the basic pp reaction (27.2) was not essential to nucleosynthesis out of hydrogen, provided one could envisage a network of reactions involving heavier nuclei as catalysts. If any of the stable isotopes of carbon, nitrogen or oxygen are mixed with the predominant hydrogen, as is the case with all but the oldest stars†, the sequence of reactions shown in figure 27.1 will be established, the net result being once more the fusion of four hydrogen nuclei to form one ^4He nucleus.

It will be seen that this process can be initiated at any point, i.e. by any of the stable isotopes of carbon, nitrogen or oxygen (shown in circles in the figure), and that eventually all the reactions will be taking place. Nevertheless, three distinct cycles, labelled 1, 2 and 3 in figure 27.1, each producing ^4He out of hydrogen, can be recognised. Cycles 1 and 2 are linked together by the two possible outcomes for $^{15}N + {}^1H$, and by the two possible ways for forming ^{14}N, while cycles 2 and 3 are linked together by the two possible outcomes of $^{17}O + {}^1H$, and by the two possible ways of producing ^{15}N. In all three cycles the formation of one ^4He nucleus involves two β^+ decays, since the basic process followed is always of the form (27.3).

Originally, only cycle 1, the so-called carbon cycle, was considered, being proposed independently by Bethe and by von Weizsäcker in 1938, at which time the right-hand branch of the $^{15}N + {}^1H$ reaction was unknown, since it is some 10^{-4} times less probable than the left-hand branch. The subsequently discovered cycles 2 and 3 do not contribute significantly to the total energy output. Although the total number of C, N and O nuclei remains

† The interstellar medium out of which new stars condense is becoming steadily richer in heavier nuclei as stars continue to spew out their contents, either in supernova explosions or otherwise.

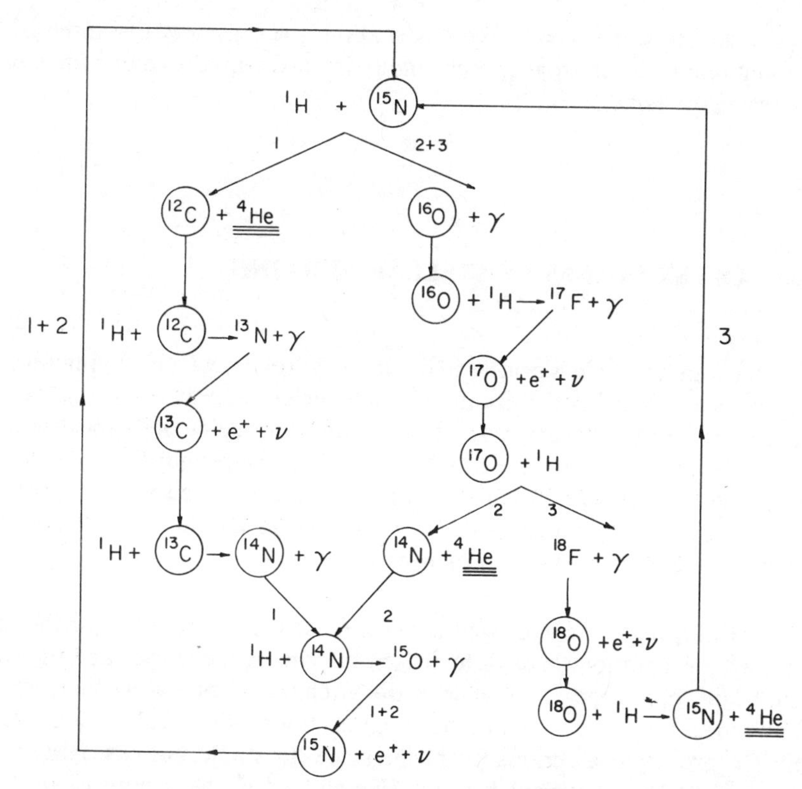

Figure 27.1 CNO tri-cycle. Catalysts (stable isotopes of C, N and O) are in the circles. Numbers 1, 2 and 3 label cycles.

unchanged during these cycles, the numbers of individual species may be changed, and in fact most of these nuclei are transformed into ^{14}N.

Relative importance of pp and CNO processes. Both processes will in principle always contribute to the synthesis of ^4He out of hydrogen. However, the Coulomb barriers that are involved in the CNO tri-cycle will be considerably higher than those involved in the various pp chains. Thus the relative importance of the CNO tri-cycle compared with the pp chains will be greater the higher the central temperature of the star (see equation (E.31) and the discussion that follows it). The sun, for example, is a relatively cool hydrogen-burning star, and the CNO tri-cycle plays a negligible role†. On the other hand, we know that the CNO process must have

† It is because of the extreme slowness of the pp process that the sun has remained essentially unchanged for more than four billion years. Now without the resulting great stability of the earth's climate it would have been impossible for complex life forms to have evolved, so that we may say that their existence depends on the extreme weakness of β decay processes.

taken place in some stars, since this is about the only way of synthesising the abundant ^{14}N (nitrogen, incidentally is an essential element for living processes on earth).

§28 LATER STAGES OF STELLAR BURNING

Having seen how ^4He is synthesised out of hydrogen, we turn to the formation of heavier nuclides. Virtually all such nuclei can only be synthesised as a result of the transmutation of ^4He, the exceptions being the stable isotopes of Li, Be and B, which were produced either in the primordial big bang or by processes such as the spallation of heavy nuclei by cosmic ray protons of very high energy (none of the $A = 7$ nuclei formed in the pp II and III chains survive the exhaustion of hydrogen).

Helium burning. As hydrogen is depleted at the centre of a star the rate of energy generation through hydrogen burning will decrease and the core will begin to contract, so that with the attendant temperature increase we may expect the onset of helium-burning reactions. Historically, the burning of ^4He presented a serious problem, since no stable nucleus exists with $A = 5$ or 8. Thus reactions between ^4He and ^1H in the regions of the star where there still remain significant amounts of hydrogen are of no avail, while the simple fusion of two ^4He nuclei according to

$$^4He + {}^4He \rightarrow {}^8Be \tag{28.1}$$

seems to be similarly fruitless, since the ^8Be nucleus simply decays back into two α particles.

We could consider the possibility that the traces of ^{12}C, for example, that are present in most stars might act once more as a catalyst. However, while the radiative capture

$$^4He + {}^{12}C \rightarrow {}^{16}O + \gamma \tag{28.2}$$

is indeed a significant reaction in helium burning, as we shall see below, there is no way in which ^{12}C can be regenerated, so that this process could serve to transmute ^4He only in an amount equal to the original amount of ^{12}C. In any case, this would still leave open the question of the origin of the ^{12}C.

An exhaustive study shows that the only possibility is that three ^4He nuclei react together:

$$3\ {}^4He \rightarrow {}^{12}C + \gamma\,. \tag{28.3}$$

That three ^{4}He nuclei could come together, and stay together long enough for the reaction (28.3) to take place, is, for the following reason, less improbable than might appear to be the case at first sight. Although ^{8}Be is indeed particle-unstable, bombardment of ^{4}He on ^{4}He reveals a narrow resonance, $\Gamma = 2.5$ eV, at a CM energy of 92 keV. This implies the formation of a compound nucleus with a lifetime of 2.6×10^{-16} s, according to equation (21.2). Now at this energy the one ^{4}He nucleus would pass by the other in a time of the order of 10^{-21} s if there were no interaction between them, so that the existence of the resonance implies that the two nuclei do stick together for what is, relatively speaking, a very long time. The capture of a second ^{4}He nucleus before the break-up of the ^{8}Be compound nucleus then becomes a much less improbable event. This second capture is also resonant, the ^{12}C being formed in an excited state of energy 7.654 MeV, which then decays radiatively to the ground state.

Helium burning through this so-called 3α reaction stabilises the star when the central temperature has risen to 10^8 K (note that hydrogen will still be burning in a shell surrounding the helium-burning core; stars at this stage of evolution have moved off the 'main sequence' and become, for example, 'red giants'). Once significant amounts of ^{12}C have been formed reaction (28.2) can begin, so that the overall result of helium burning is the formation of ^{12}C and ^{16}O, the relative amounts depending very much on the prevailing conditions (the calculations are subject to considerable uncertainty on account of the inadequacy of data concerning reaction (28.2)).

At the same time the helium can initiate the following very important sequence of processes with any ^{14}N that may be present (this, it will be recalled, is the principal by-product of the CNO burning of hydrogen)

$$^{14}\text{N} + {}^4\text{He} \to {}^{18}\text{F} \tag{28.4}$$

$$^{18}\text{F} \to {}^{18}\text{O} + \text{e}^+ + \nu\,. \tag{28.5}$$

$$^{18}\text{O} + {}^4\text{He} \to {}^{22}\text{Ne} \tag{28.6}$$

$$^{22}\text{Ne} + {}^4\text{He} \to {}^{25}\text{Mg} + \text{n}\,. \tag{28.7}$$

This sequence does not contribute significantly to energy generation, but the neutrons produced in the last reaction can be radiatively captured to form new nuclides, especially heavy ones, i.e. nuclides with $A > 56$. This is the so-called *s process* ('s' meaning 'slow'), which will be discussed in more detail in §29.

Carbon and oxygen burning. Eventually the helium content at the centre of a star will fall to the point where helium burning is no longer sufficient to maintain the stability of the star. Thus the core contracts once more and the temperature rises until burning of the ^{12}C formed from helium begins at around 7×10^8 K. As seen in equation (19.15), there are several possible reaction modes for the $^{12}C + ^{12}C$ system, but the most abundant product of this phase is ^{20}Ne

$$^{12}C + ^{12}C \rightarrow ^{20}Ne + ^{4}He. \tag{28.8}$$

As ^{12}C is depleted the temperature of the core will rise again, but by the time it has reached the point where significant oxygen burning can occur, at a temperature of 10^9 K (note, incidentally, the increasing sensitivity to temperature with increasing Z, as discussed in Appendix E), very little ^{12}C will remain. Thus the reactions between ^{12}C and ^{16}O are of little interest, and all that matters is the reaction of ^{16}O with itself, the principal product of which is ^{28}Si

$$^{16}O + ^{16}O \rightarrow ^{28}Si + ^{4}He\,. \tag{28.9}$$

At the end of oxygen burning one might naively expect that as the temperature rose again fusion would begin between the relatively heavy nuclei now present. In particular, we might look for fusion of ^{28}Si with itself to form ^{56}Ni, which would then undergo two β^+ decays to ^{56}Fe: this latter, being the most strongly bound of all nuclei, must be the end product of all processes in which thermal equilibrium is maintained. However, before the very high temperatures necessary for this to take place are reached a new process sets in, and ^{56}Fe is formed by a more circuitous route, as we now discuss.

Photodisintegration. The interior of a star in thermal equilibrium resembles closely a cavity (in the thermodynamical sense) and must therefore be filled with electromagnetic radiation whose spectral form is determined by the temperature according to the Planck black-body radiation law, thus

$$n(\omega)\,d\omega = \frac{\omega^2\,d\omega}{c^3\pi^2[\exp(\hbar\omega/kT) - 1]} \tag{28.10}$$

in which $n(\omega)d\omega$ represents the number of photons per unit volume whose frequency lies in the interval ω to $\omega + d\omega$. At the temperature of the surface of a star like the sun, around 6000 K, the maximum of this distribution is in the visible region, and the number of photons whose energy corresponds to that of nuclear γ rays is quite negligible.

On the other hand, at temperatures of the order of 10^9 K, as in the oxygen-burning regions of a star, there will be a significant flux of photons energetic enough to cause photodisintegration of the nuclei present. For example, the reaction

$$^{28}Si + \gamma \rightarrow ^{24}Mg + ^{4}He \tag{28.11}$$

is possible, along with the photo-ejection of protons and neutrons from the ^{28}Si. Of course, the inverse reaction to (28.11) is also possible, and one might wonder why the $^{24}Mg + ^{4}He$ system does not go over completely into the more strongly bound ground state of ^{28}Si. But it must be remembered that according to statistical mechanics any system in thermal equilibrium at a finite temperature T may find itself in an excited state of energy E above the ground state, with a probability that is determined mainly by the Boltzmann factor, $\exp(-E/kT)$ (there are also phase-space factors). Now the system $^{24}Mg + ^{4}He$ may be regarded as an excited state of ^{28}Si with an excitation energy of the order of S_α, the separation energy of ^{4}He from ^{28}Si. Thus when thermal equilibrium† at a temperature T prevails in a sample of ^{28}Si, we may expect a fraction of the order of $\exp(-S_\alpha/kT)$ to be decomposed into $^{24}Mg + ^{4}He$. Likewise, a certain fraction will be decomposed into $^{27}Si + n$ and $^{27}Al + p$. (For the same reason, at these high temperatures a significant fraction of all nuclei will be in particle-stable excited states.)

The actual situation is complicated enormously by virtue of the fact that the photonuclear disintegration products of one ^{28}Si nucleus can be captured by other ^{28}Si nuclei, and also by nuclei of any other species that are present. Such captures will actually be energetically favoured if the separation energies of the captured particles in their new 'host' nuclei are greater than in ^{28}Si: the capture will not be complete, of course, but the Boltzmann factor will be smaller than it was before. And since there is a general tendency for nuclei to be more strongly bound the closer the mass number is to 56, it follows that the overall effect will be for the lighter nuclei to break up and their fragments fuse with the heavier nuclei—the light nuclei become lighter and the heavy nuclei heavier‡. It is by this so-called *photonuclear rearrangement* that the gas consisting of almost pure ^{28}Si drifts over into nuclei of mass 56: one speaks of 'silicon melting'. Of course, at some point β^+ decays will intervene, bringing us to the end point of ^{56}Fe.

A similar process occurs on a much smaller scale between the carbon- and oxygen-burning phases, the ^{20}Ne formed during the former phase being transformed into ^{24}Mg by the following sequence

$$^{20}Ne + \gamma \rightarrow ^{16}O + ^{4}He \tag{28.12}$$

$$^{4}He + ^{20}Ne \rightarrow ^{24}Mg\ . \tag{28.13}$$

† Actually, it must not be supposed that there is always enough time for a complete thermal equilibrium to be established during this phase.

‡ It must not be thought that this process has to be energetically favoured at each step. For example, ^{28}Si itself constitutes a local 'island of stability', and removal of either a proton, a neutron, or an α particle from one ^{28}Si nucleus with subsequent attachment to another ^{28}Si nucleus is energetically unfavoured. Nevertheless, the process will not be impossible, because the Boltzmann factor will still be finite, and eventually nuclei so much heavier than ^{28}Si will be formed that subsequent photonuclear rearrangement will be energetically favoured. It remains true that ^{28}Si constitutes a 'bottle-neck': this accounts for its great abundance.

Termination of thermonuclear evolution. Because the star contracts between successive phases of nuclear burning the density of free electrons at the centre of the star increases and, as explained in Appendix D, they may become degenerate. When this happens it is possible that the associated electron pressure will be sufficient to stabilise the star as a *white dwarf*, provided it is not too heavy (in Appendix D we give a crude estimate of 1.4 solar masses for this upper limit). Further contraction, and hence further temperature increases, will be impossible, so that nuclear evolution will be brought to a standstill, even if the ^{56}Fe limit has not been reached.

For heavier stars the degeneracy pressure will never be sufficient to 'freeze' the star in this way, and the core of the star must inevitably evolve towards ^{56}Fe. However, even though ^{56}Fe is the most stable nucleus, it must not be concluded that once the heavy star has reached this point it will finally have acquired some stability. On the contrary, just because the star has reached the minimum in the binding energy curve of figure 14.2 no further source of nuclear energy is available, and the star has no choice but to continue to contract, drawing from gravity the energy it must radiate.

In fact, the problem is aggravated by the fact that as the star contracts the Fermi energy of the degenerate electrons increases until it becomes energetically favourable for them to be captured through inverse β decay by nuclear-bound protons, transforming them into neutrons, and emitting neutrinos in the process (see §§11 and 16 and Appendix D). The removal of electrons results in a reduction of the pressure they exert, thereby accelerating the contraction. (The loss of electron pressure is not compensated by the neutrinos that are created, since these hardly interact at all with the cooler envelope of the star, and they simply escape from the star, carrying with them a significant amount of the star's energy.)

A further source of instability, also tending to accelerate the contraction, arises from the photodisintegration of the ^{56}Fe that sets in with rising temperature† (because the Boltzmann factor is always finite the core never consists of pure ^{56}Fe, more and more breaking up as the temperature rises). Since this process is highly endothermic it draws off a lot of the gravitational energy that otherwise would have gone to raising the temperature of the star still further and would thereby have tended to slow down the contraction.

Supernova explosion. As a result of all these various processes, the ever-accelerating contraction turns into a catastrophic implosion, which continues until all the nuclei of the star's core actually touch each other and fuse into one enormous nucleus. Actually, because nearly all the free

† We see that there is a somewhat elusive quality to the ^{56}Fe end point. It cannot be reached at low temperatures since the reaction rates are then too slow, while at high temperatures the equilibrium configuration no longer consists of pure ^{56}Fe.

electrons will by then have been captured, this super-nucleus into which the core of the star will have collapsed consists almost entirely of neutrons.

The formation of this core, which may be regarded as a very large piece of nuclear matter (§§13 and 15), will lead to a very significant and sudden braking of the collapse, both because of the short-range repulsion of the nucleon–nucleon forces and the fact that the nucleons themselves now constitute a degenerate gas (we may say that nuclear matter is much more incompressible than ordinary matter). This sudden braking of the core's collapse often triggers off, in ways that are not completely clear, a *supernova explosion*, in which the entire envelope of the star, containing by far the greater part of its mass, is blown off into space. Some idea of the scale of this catastrophic release of energy can be had from the fact that the luminosity of a supernova can be as high as 1% of that of an entire galaxy; the supernova explosion in our own galaxy that was observed by Chinese astronomers in 1054 AD was visible in daylight for one month. (The aftermath of that explosion can still be seen today: the Crab Nebula marking the expanding cloud of debris from the shattered envelope.)

If the residual neutron core is not too massive (see Appendix D) it will stabilise as a *neutron star*, observable as a *pulsar* (such an object can be seen today at the centre of the Crab Nebula). But if the mass of the core exceeds a limiting value of the order of a couple of solar masses neither the nucleon-degeneracy pressure nor the nucleon–nucleon forces will be able to resist the gravitational compression, and the original collapse will continue until all matter disappears into a *black hole* (Appendix D).

Fascinating though this question of the core's fate may be, it has no bearing on the origin of the elements and of the abundances of the various nuclides observed in the solar system, and in the galaxy at large. For this we must turn rather to the envelope of the exploding star. It must be realised that at the instant of the explosion successive layers of the envelope further and further away from the core will have reached less and less advanced stages of thermonuclear evolution: all the different stages of hydrostatic burning that we sketched in the foregoing, along with the associated s-process neutron-induced synthesis (see the next section), will be going on somewhere or other in the envelope. Thus the result of the explosion is to blow out into space quantities of *all* the various nuclides† that have appeared at any time during the star's history: nuclides that long ago vanished at the core of the star will reappear later nearer the surface.

But that is not all: if it were then many naturally occurring nuclides would never have the opportunity of being synthesised (in particular, no element heavier than bismuth could be formed), and in other cases the calculated abundances would be quite wrong. The fact is, that after nucleosynthesis

† This includes many nuclides with $A > 56$, formed through the s process, as described in the next section.

has been going on steadily in the hydrostatically stable star over a period of hundreds of millions of years, a significant component of the overall synthesis corresponding to the observed nuclidic abundances occurs in the last few seconds of the star's life, during the supernova explosion†.

Just as in the hydrostatic era, both charged-particle and neutron reactions will take place, and each plays a vital role in nucleosynthesis. The latter reactions occur as a result of a very intense neutron flux that is associated with the explosion‡, and constitute the so-called *r process* of nucleosynthesis ('r' meaning 'rapid') to distinguish it from the s process in hydrostatic stars. Here too the main result is the synthesis of heavy nuclei, $A > 56$, but not the same as those formed by the s process; both processes will be discussed in the next section.

By way of contrast, the main influence of *explosive burning*, as the charged-particle reactions in the exploding star are referred to collectively, is on lighter nuclei: even in these violent conditions there is little significant penetration of the Coulomb barrier of nuclei with $A > 65$ or so. We shall not even attempt to summarise this extremely complex process, beyond saying that not only does it synthesise some nuclides that cannot be formed by the hydrostatic star, but that it can also destroy some of the products of hydrostatic burning.

To conclude these remarks on supernovae, it should be pointed out that it is not only through explosions of this sort that the products of stellar nucleosynthesis are added to the interstellar medium. There are a variety of much less violent processes that can expel matter from a star before it has reached the point of exploding as a supernova. Thus the composition of the interstellar medium is not determined solely by stars that have reached the supernova stage, but by less well evolved stars also.

§29 NEUTRON-CAPTURE NUCLEOSYNTHESIS

We pointed out at the end of the previous section that because of the Coulomb barrier there is very little chance of having any charged-particle reactions for nuclei with $A > 65$, even under the extremely energetic conditions of a supernova explosion. The question then arises as to how the heavy nuclei were formed. In fact, the discovery of technetium in the spectra of hydrostatically stable stars shows that heavy elements are being synthesised

† It is remarkable that *all* fissile and fertile nuclei are synthesised in these last few seconds.

‡ It is far from clear whether this neutron flux comes from the neutronisation of the collapsing core, or from (α, n) reactions in helium-burning shells of the envelope.

even before the supernova stage is reached: since there exists no isotope of this element with a half-life longer than a few millions years (^{97}Tc)† the observed nuclei could not possibly have been present in the galactic dust out of which the star condensed.

We recall now that neutron-induced reactions, unlike charged-particle reactions, are not inhibited by any Coulomb barrier, and in fact, because of the $1/v$ law (see §20), the lower the energy of a neutron the greater the cross section for its capture. Now if a particular nucleus is exposed to a flux of neutrons it will be able to radiatively capture not just one neutron but a whole succession of neutrons, progressively heavier isotopes of the same chemical element being formed in the process. Eventually, the resulting β^- instability will carry the nucleus over into an isobar with Z value increased by unity, so that a new chemical element, with a higher place in the periodic table, will have been formed. The process can then begin again, if there is still a neutron flux, although it will eventually be brought to an end by particle emission or fission when very heavy nuclei are reached.

It turns out that there really is no alternative to some such combination of sequential neutron capture and β^- decay for the synthesis of nuclei with $A > 56$, and we are thus left with the problem of identifying suitable sources of neutrons within stars—remember that the neutron is an unstable particle (half-life 11 minutes) so that the neutrons must be produced at essentially the same time as they are being radiatively captured. We have already tentatively identified two quite different sources of neutrons, one in hydrostatically stable stars and the other in supernovae, giving rise to the so-called s and r processes, respectively. The existence of technetium in certain stellar atmospheres shows that the s process must be occurring to some extent, while its inability to explain the abundances of *all* the heavy nuclides shows that the r process must also be playing a significant role. We now examine these two processes in more detail.

The s process. We have seen in §28 how neutrons can be produced during the helium-burning phase of a hydrostatically stable star. Now the flux of neutrons produced in this way is so weak that many years will elapse between successive neutron captures by a given nucleus. Thus any β^- unstable nucleus that is formed by neutron capture will undergo β^- decay before it can capture another neutron: the characteristic 'slowness' of this process is that of neutron capture compared with β decay.

We see then that increasingly heavy isotopes of a given element can be synthesised in this way provided *all* are stable, but any unstable isotope will bring the sequence to a halt: the s process will not be able to synthesise isotopes lying beyond such a gap, even if they are stable. For example,

† Actually, this isotope is by-passed by the s process, and the longest lived technetium isotope that can be so formed is ^{99}Tc, with $t_{1/2} = 2 \times 10^5$ years.

referring to figure 29.1, we see that the heaviest Cd isotope that the s process can synthesise is ^{114}Cd, even though ^{116}Cd is stable. Likewise, ^{122}Sn and ^{124}Sn cannot be synthesised by the s process, even though they too are stable.

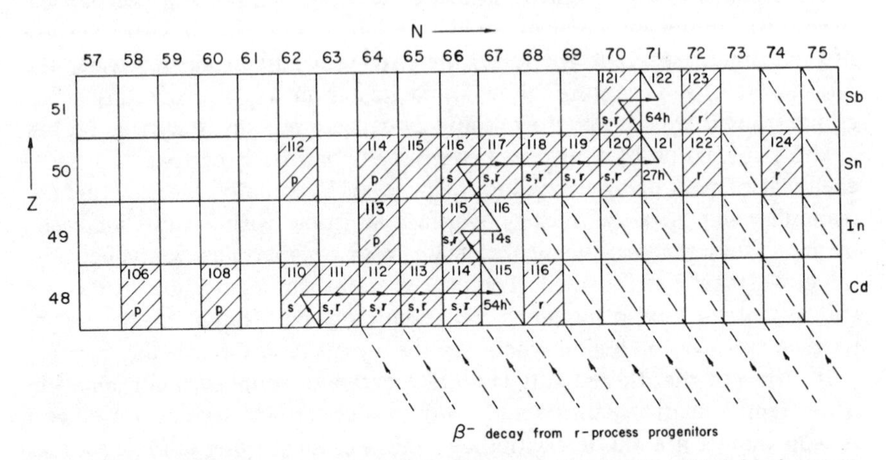

Figure 29.1 Neutron-capture synthesis of elements $Z = 48–51$. All stable isotopes are shown (hatched). The only β unstable nuclei shown are the ones that terminate the sequence of isotopes of given Z. The full line shows the path of the s process.

On the other hand, the β^- decay that terminates the sequence of isotopes of a given element leads to the generation of a new element, with Z increased by unity, at which point the whole sequence of events can begin again with the new element. It must be realised that just as the heaviest stable isotopes of an element may not be formed by the s process, it is also possible that some of the lightest isotopes will not be formed either: for example, it is clear from figure 29.1 that ^{106}Cd, ^{108}Cd, ^{113}In, ^{112}Sn, ^{114}Sn, and ^{115}Sn cannot be synthesised by the s process.

In any case, the s process will bring us zigzagging along the floor of the valley of stability towards heavier and heavier elements. The heaviest nucleus that can be formed in this way is ^{209}Bi: the process is brought to a halt by α decay in the following sequence of events

$$^{209}\text{Bi} + \text{n} \rightarrow {}^{210}\text{Bi} \xrightarrow[5\text{ d}]{\beta^-} {}^{210}\text{Po} \xrightarrow[138\text{ d}]{\alpha} {}^{206}\text{Pb}$$

$$^{206}\text{Pb} + \text{n} \rightarrow {}^{207}\text{Pb}$$

$$^{207}\text{Pb} + \text{n} \rightarrow {}^{208}\text{Pb}$$

$$^{208}\text{Pb} + \text{n} \rightarrow {}^{209}\text{Pb} \xrightarrow[3.1\text{ hr}]{\beta^-} {}^{209}\text{Bi}.$$

The zigzag path is in effect terminated by a loop.

Somewhat more problematical than the end point of the s process is the question of its starting point. In principle, it could begin independently on any nucleus present in the star. Now, except in the case of a new galaxy, the interstellar gas out of which a star is born will contain at least traces of all stable nuclides, but beyond $A = 56$ the abundances fall off rapidly. On the other hand, the capture cross section for thermal neutrons†, averaged across the Maxwell–Boltzmann distribution, generally increases with A, essentially because the level density in the compound nucleus increases (note that at these temperatures thermal energies are of the order of keV, and the Maxwell–Boltzmann distribution of energy will cover many levels of the compound nucleus, at least for heavy nuclei). For this reason we are left with nuclei in the vicinity of $A = 56$ as the most fruitful seeds for the s process. It is nevertheless interesting to note that heavy nuclei which do not lie on the path of the 'primary' s process could serve as seeds leading to new branches, e.g. starting with ^{114}Sn the s process could yield ^{115}Sn (see figure 29.1).

Finally, we should point out that the greater the neutron-capture cross section of a nucleus the greater the rate with which this nucleus will be transformed into a heavier nucleus, and the lower the abundance with which it is synthesised in the s process. Thus we should expect an inverse correlation between the galactic abundance of a nuclide and its neutron-capture cross section. This is precisely what is found to be the case in reality. In particular, nuclei with N or Z as magic numbers are found to be exceptionally abundant, such nuclei having abnormally low neutron-capture cross sections, essentially because of the stability of the closed-shell structure.

The r process. We have already pointed out that there are many heavy nuclei that cannot be synthesised by the s process. In particular, no element beyond bismuth ($Z = 83$) can be formed in this way, and even for lighter elements there are many isotopes that are missed by the path of the s process (see figure 29.1).

At the same time we have also suggested that it is not only in hydrostatically stable stars, the site of the s process, that there is a flux of neutrons: we may also expect a strong, albeit brief, neutron flux during a supernova explosion. A completely different mode of neutron-capture nucleosynthesis, the so-called r process, can now be envisaged, if it is assumed that this supernova flux of neutrons is so strong that the rate of successive neutron captures by any given nucleus will be extremely *rapid* compared with β decay rates. In that case the sequence of neutron captures by a nucleus of given Z will not come to a halt at the first β unstable isotope encountered. Rather, any such gap will be jumped, and capture of neutrons will continue right into the region of very neutron-rich nuclei, close to the neutron drip line. Actually, because of the very high temperature prevailing

†The neutrons produced in reactions such as (28.7) are rapidly thermalised.

in the supernova explosion, capture of neutrons for nuclei with given Z will stop at some point short of the drip line, since some nuclei will not be in their ground state, and neutrons will be able to leak out as fast as they are captured, even though the ground state may be stable with respect to neutron emission. Essentially, it is the equilibrium between the (n, γ) reaction and its inverse (γ, n), that characterises the end point of neutron captures for a given value of Z, and it is clear that the position of this end point will depend not only on the temperature but also on the neutron flux, N being increased by lower temperatures and higher neutron fluxes.

Once this end point, or *waiting point*, is reached the process will wait until the inevitable intervention of a β decay, which will carry the waiting point nucleus over into an isobar with Z larger by unity. This nucleus will still be very neutron-rich, but because of the higher value of Z the waiting-point value of N will in general be several units higher than before. Thus after the β decay neutron capture will resume at once until the next waiting point is reached.

In this way, starting from seed nuclei in the region of ^{56}Fe, progressively heavier elements will be synthesised. The process is in this respect somewhat similar to the s process, but there is this fundamental difference that the r process path, which is the locus of the waiting points, rides high on the neutron-rich side of the valley of stability, instead of following the stability line along the valley bottom. Furthermore, the timescale is incomparably more rapid, being of the order of seconds.

The path of the r process does not come to an end with the formation of bismuth, as is the case with the s process, but continues into the transuranic region, where it is eventually terminated by fission. It has already been pointed out (§§14, 17 and 18) that spontaneous fission sets an upper limit on the mass of nuclei that can have any sort of existence, but in the r process the fission rate is enhanced enormously by the neutron flux: it is *neutron-induced* fission that brings the r process to an end.

Actually, the nuclei that undergo fission in this way are not lost to the r process. On the contrary, the two heavy nuclei that result from the fission of one nucleus will serve as seeds for a new r process chain, and in fact it is easy to see that a cycle will be set up and repeated as long as the neutron supply holds out, with the number of heavy nuclei being doubled by fission at the end of each such cycle.

However, at the end of the blast of neutrons accompanying the supernova explosion all nuclei on the r process chain will drift over towards the stability line by virtue of β^- decay. The ultimate nucleus formed by a given progenitor on the r process path will be the most neutron-rich stable isobar with the given value of A. This process is represented by the broken arrows in figure 29.1, with the stable nuclei so formed being labelled 'r'. (Actually, the situation will be complicated by the occurrence not only of β delayed neutron emission but also of the analogous process of β delayed fission.)

It will be seen from this figure that many nuclei that can be formed by the s process can also be formed by the r process: they are labelled 's, r'. However, since only one stable nucleus for each value of A can be formed by the r process, it follows that some nuclei can be formed only by the s process, e.g. ^{116}Sn, which is 'shielded' from the r process by ^{116}Cd. At the same time many neutron-rich stable nuclei can be formed only by the r process.

The abundances of the r process stable nuclei will clearly have nothing to do with their neutron-capture cross sections. Rather, they will be determined by the abundances of their respective isobaric progenitors on the r process path. These latter will depend in turn on several properties of the nuclei involved, in particular on their binding energies, since these determine not only the positions of the waiting points for given temperature and neutron flux, but also the rates of the β decays at the waiting points (through their Q values). Now all the relevant nuclei, being highly neutron-rich, are very unstable and it is not possible to measure their masses in the laboratory. Thus recourse has to be made to the semi-empirical mass formula, and it is this application to the astrophysical r process that provides the main motivation for the continuing efforts that are being made to refine it beyond the rudimentary form discussed in §14.

The p process. It will be seen from figure 29.1 that there are a few heavy nuclei ($A > 56$) whose formation can be accounted for by neither the s nor the r process. Altogether there are 30 or so nuclei of this type and since they are all richer in protons than the s or r nuclei they are known as *p-process nuclei*, although, as we shall see, more than one mechanism may be responsible. But however they are produced it is fairly clear that they are derived by secondary processes from s-type or r-type nuclei, since they are much less abundant than the latter.

The following p processes have been proposed:

(*a*) (p, γ) and (p,n) reactions. Reactions between hydrogen and heavy nuclei could take place during explosive burning.

(*b*) (γ, n) reactions. Photons of sufficiently high energy will be present as part of the black-body background for $T > 10^9$ K.

(*c*) For these temperatures a certain number of (e^+ e^-) pairs will be in thermal equilibrium with the black-body photon spectrum. Capture of these free positrons could be responsible for some of the p nuclei.

§30 ARTIFICIAL THERMONUCLEAR FUSION

We have on a number of occasions compared and contrasted the fission and fusion processes as macroscopic sources of energy. One respect in which they differ is that whereas the fission chain reaction was first realised artificially before being recognised as a possible natural phenomenon, the contrary is the case with fusion. Furthermore, whereas the artificial fission chain reaction was first established on a controlled scale before being realised explosively, artifical fusion has at the time of writing only been achieved as an explosion: a self-sustaining, controlled artifical fusion process is still only a dream, albeit one whose fulfilment is being sought most strenuously.

One of the most important considerations for artificial thermonuclear fusion is the ignition temperature, i.e. the temperature which must first be established before the fusion reaction can become self-sustaining. Generally speaking this will be lower the lighter the reacting nuclei, although the presence of resonances may modify this trend imposed by the Coulomb barrier.

A further consideration that limits drastically the choice of reaction is that, unlike the case of the stars, where time is no object, the intervention of a β decay would render artificial fusion intolerably slow. Thus hydrogen, ^1H, can be of no possible interest for artificial fusion, and one turns therefore to deuterium, which reacts with itself thus

$$^2H + {}^2H \rightarrow \begin{cases} ^3He + n + 3.3 \text{ MeV} \\ ^3H + {}^1H + 4.0 \text{ MeV} \\ ^4He + 23.8 \text{ MeV}. \end{cases} \tag{30.1}$$

Unfortunately, the most energetic of these modes, the last one, has a very small cross section compared with the two others, which have roughly equal cross sections. On the other hand, the ^3He and ^3H produced in the first two modes can each react with the remaining deuterium, thus

$$^3He + {}^2H \rightarrow {}^4He + {}^1H + 18.3 \text{ MeV} \tag{30.2}$$

$$^3H + {}^2H \rightarrow {}^4He + n + 17.6 \text{ MeV}\,. \tag{30.3}$$

Finally, the protons and neutrons formed in these reactions could combine to regenerate deuterium

$$^1H + n \rightarrow {}^2H + 2.2 \text{ MeV}\,. \tag{30.4}$$

Thus the overall process is equivalent to

$$4\,{}^2H \rightarrow 2\,{}^4He + 47.7 \text{ MeV}\,. \tag{30.5}$$

This represents an energy release of almost 6 MeV per nucleon, compared with less than 1 MeV per nucleon in the case of fission. It follows that if this process could be harnessed some 10^{10} joules could be extracted from one litre of *ordinary* water (this contains about one ^{2}H atom for every 6500 ^{1}H atoms), which is about 300 times more energy than is obtained by the combustion of the same volume of gasoline. When it is realised how much water there is in the oceans we see that we have here an energy supply that will certainly outlive the human race.

The ignition temperature for the primary reaction (30.1) is around 4×10^{8} K: this is the temperature that must first be established by other means before the reaction can become self-sustaining. There are several ways in which very high temperatures of this order can be sought. In the hydrogen bomb (see below), for example, the necessary temperatures are invariably achieved through the prior detonation of a fission bomb, which serves effectively as trigger. In the case of controlled fusion, this recourse to a fission bomb will not of course be possible, and one must look for some other means. An enormous amount of effort is being expended at the present time on exploring two quite different approaches to the problem. In one of these, of which there exist many variants, the reacting nuclei are in the form of a plasma, i.e. ionised gas, which may be heated by a suitable application of magnetic fields; with an appropriate configuration the heated gas can be kept away from the walls of the container. In the other approach the reacting nuclei are in the form of pellets which are subjected to extreme pressure through bombardment by laser beams.

At the present time the ignition temperature for the reaction (30.1) is still a long way from being achieved. On the other hand it is known that the ignition temperature for reaction (30.3), between deuterium and tritium, is about 4×10^{7} K, considerably lower than for the deuterium–deuterium reaction (30.1), essentially because of a resonance in the cross section. Thus it is quite clear that the first fusion reactors will be fuelled by a D–T mixture, and will function exclusively on the reaction (30.3).

While the world reserves of deuterium are, as we have seen, effectively unlimited, there arises now the question of the supply of tritium. This, of course, does not occur naturally, since it is unstable, with a half-life of 12 years. However, as the tritium stock is consumed in a reactor of this sort it can be replenished automatically by surrounding the reactor with a blanket of lithium, the neutrons released in the reaction (30.3) then leading to the reaction

$$n + {}^{6}\mathrm{Li} \rightarrow {}^{4}\mathrm{He} + {}^{3}\mathrm{H}\ . \qquad (30.6)$$

In fact, to maintain the tritium stock in this way it would be necessary for every emitted neutron to be captured in the lithium blanket, which is clearly impossible in practice. Fortunately, natural lithium contains not only ^{6}Li, but also ^{7}Li, in which the reaction

$$n + {}^{7}\mathrm{Li} \rightarrow {}^{6}\mathrm{Li} + 2n - 7.2\ \mathrm{MeV} \qquad (30.7)$$

takes place, the neutrons produced in the reaction (30.3) having 14 MeV energy. There is thus an amplification of neutrons, and in fact more tritium will be generated than is consumed.

We see now that it is the world's reserves of lithium that impose the essential limitation on the amount of energy that can be generated by the D–T fusion process; it is estimated that this is of the same order as the available energy in the world's fossil fuel reserves. It follows that mankind's ultimate hopes for energy supplies must lie with the D–D process.

However, the realisation of this hope must lie far in the future, since it is clear that even the more modest D–T process has no chance of being commercially viable before the twenty-first century. At the present time laboratory research has reached the point where the D–T reaction (30.3) can indeed be made to take place at an observable rate in a magnetically confined plasma, although the energy output is still lower than the energy input necessary to produce the required heating, i.e. the 'break-even' point has not yet been reached.

Fission–fusion hybrid. Although the attainment of the 'break-even' point in a magnetically confined D–T plasma may still be some years away, even on a laboratory scale, the fact that the reaction (30.3) has already been observed in a plasma means that there is a much more immediate prospect of this fusion process serving as a copious source of neutrons, even though a net amount of energy would have to be supplied. It has now been suggested† that such a large-scale neutron source could be used to convert fertile material to fissile fuel, i.e. ^{238}U to ^{239}Pu, or ^{232}Th to ^{233}U. The proposal is, therefore, that D–T reactors be surrounded by a blanket of natural uranium or thorium, the fissile material generated being extracted for use in thermal reactors. There would then be a net energy gain, even if the D–T reactor itself had not yet reached the 'break-even' point. Such a hybrid process, which should make it possible to burn essentially all the earth's fertile material (see p. 166), offers an attractive alternative to breeding in a fission reactor: we have spoken in §25 of the dangers of the fast breeder, essential for the breeding of ^{239}Pu from ^{238}U, and the fact that it is not clear whether it is possible to breed ^{233}U under any circumstances.

Hydrogen bomb. This uses essentially the D–T reaction (30.3). Since the deuterium and tritium must both be in a highly dense state in order for the reaction to proceed rapidly, the world's first thermonuclear explosion, detonated by the USA at Bikini in November 1952, used liquefied deuterium and tritium. However, the necessary refrigerating equipment made the 'device' quite unpractical as a bomb (it weighed 65 tonnes). Moreover, the relatively short half-life of tritium (12 years) makes its use in a bomb rather inconvenient.

† See H Bethe 1979 The fusion hybrid *Physics Today* **32** (5) 44.

Both the problem of high density and that of tritium instability are solved in actual bombs by using as explosive the solid lithium hydride, in the form of ^{6}LiD. Under the blast of neutrons from the triggering fission bomb tritium is generated from the ^{6}Li by the reaction (30.6), whereupon the D–T reaction (30.3) can ensue under the high temperature produced by the fission bomb: note the dual role of the latter. The first such bomb was tested by the USSR in August 1953.

Usually, hydrogen bombs are made with an outer blanket of ^{238}U, which undergoes fission under bombardment by the 14 MeV neutrons produced in the reaction (30.3). In this way a tremendous amplification of the explosive force is achieved very cheaply; such a bomb is referred to as a 'fission–fusion–fission' device. On the other hand, the so-called 'neutron bomb' consists simply of a hydrogen bomb *without* the blanket of ^{238}U.

BIBLIOGRAPHY

Audouze J and Vauclair S 1980 *An Introduction to Nuclear Astrophysics* (Dordrecht: Reidel)

Barnes C A, Clayton D D and Schramm D N (ed) 1982 *Essays in Nuclear Astrophysics* (Cambridge: Cambridge University Press)

Clayton D D 1968 *Principles of Stellar Evolution and Nucleosynthesis* (New York: McGraw-Hill)

Clayton D D and Woosley S E 1974 Thermonuclear astrophysics *Rev. Mod. Phys.* **46** 755

Fowler W A 1967 *Nuclear Astrophysics* (Philadelphia: American Philosophical Society)

Rolfs C and Trautvetter H P 1978 Experimental nuclear astrophysics *Ann. Rev. Nucl. Sci.* **28** 115

Romer R H 1976 *Energy, an Introduction to Physics* (San Francisco: W H Freeman) (specifically for §30)

Trimble V 1975 The origin and abundance of the chemical elements *Rev. Mod. Phys.* **47** 877

Truran J W 1984 Nucleosynthesis *Ann. Rev. Nucl. Sci.* **34** 53

APPENDICES

APPENDIX A TRANSFORMATION OF COORDINATES FOR A TWO-BODY SYSTEM

Consider the typical laboratory experiment in which projectile a bombards target A, producing two new particles b and B, as shown in figure A.1(*a*)

$$a + A \rightarrow b + B\,. \tag{A.1}$$

In the special case of elastic scattering we shall have a ≡ b, A ≡ B. (As discussed in §19, sometimes more than two particles result from the collision, but we shall not consider such cases here. Even so, the present treatment will still be applicable to the initial pair of particles.) Our treatment is to be specifically non-relativistic, so that in general we shall not be able to consider cases where photons are involved.

We denote the respective masses of the nuclei a, A, b and B by M_a, M_A, M_b and M_B, and their respective instantaneous velocities by v_a, v_A, v_b and v_B. The asymptotic velocities, measured beyond the range of any mutual interaction, are written as u_a, u_A, u_b and u_B, respectively: the first two represent initial velocities (in the usual laboratory experiment we shall have $u_A = 0$, of course), and the last two final velocities (in the special case of elastic scattering u_b and u_B represent the final velocities of particles a and A respectively). The two particles b and B emerge at angles θ_b and θ_B, respectively, with respect to the beam direction u_a.

The total kinetic energy at any instant before collision is

$$T = \tfrac{1}{2}M_a v_a^2 + \tfrac{1}{2}M_A v_A^2 \tag{A.2a}$$

while after collision we shall have

$$T' = \tfrac{1}{2}M_b v_b^2 + \tfrac{1}{2}M_B v_B^2\,. \tag{A.2b}$$

Both of these will vary, of course, as the distance between the particles varies. The total conserved energies will be given by the asymptotic velocities:

$$E = \tfrac{1}{2} M_a u_a^2 + \tfrac{1}{2} M_A u_A^2 \tag{A.3a}$$

and

$$E' = \tfrac{1}{2} M_b u_b^2 + \tfrac{1}{2} M_B u_B^2 . \tag{A.3b}$$

The two are related by

$$E' = E + Q \tag{A.4}$$

where from the Einstein energy–mass relation (5.6)

$$Q = (M_a + M_A - M_b - M_B)c^2 \tag{A.5}$$

(this is zero for elastic scattering).

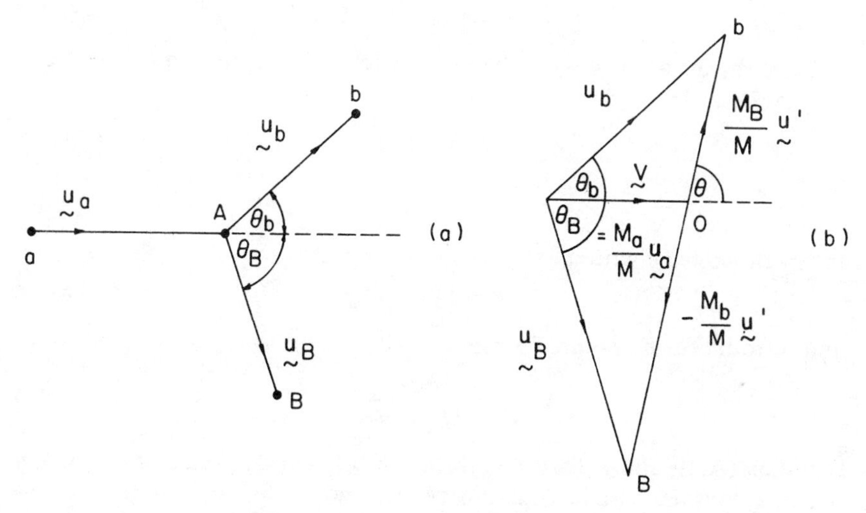

Figure A.1 Transformation of coordinates for a two-body system.

We now show that the problem of the motion of either two-body system, i.e. the one pertaining before the collision, or the one after, can be reduced effectively to the problem of the motion of a one-body system. We denote the two particles of either system by 1 and 2, and their positions in a laboratory-fixed system at any instant by $\boldsymbol{r}_1$ and $\boldsymbol{r}_2$, respectively. The equations of motion of the two particles are

$$M_1 \ddot{\boldsymbol{r}}_1 = \boldsymbol{F}_1 \tag{A.6a}$$

$$M_2 \ddot{\boldsymbol{r}}_2 = \boldsymbol{F}_2 \tag{A.6b}$$

where $\boldsymbol{F}_1$ is the force acting on particle 1, and likewise for particle 2. We assume now that Newton's third law holds

$$F_1 + F_2 = 0 . \tag{A.7}$$

(This presupposes that no external forces are acting, i.e. that the force on each particle arises solely from interactions with the other particle. Furthermore, when these interactions contain electromagnetic terms the velocities must not be so high that there is significant photon production.) Then

$$M_1\ddot{r}_1 + M_2\ddot{r}_2 = 0 . \tag{A.8}$$

We introduce now the variables

$$R = \frac{M_1 r_1 + M_2 r_2}{M_1 + M_2} \tag{A.9a}$$

which is the coordinate of the centre of mass, and

$$r = r_1 - r_2 \tag{A.9b}$$

which is the distance between the two particles. Then from equations (A.6) and (A.8) we find

$$\ddot{R} = 0 \tag{A.10}$$

and

$$\mu\ddot{r} = F \tag{A.11}$$

in which we have written

$$F \equiv F_1 = -F_2 \tag{A.12}$$

and introduced the *reduced mass*

$$\mu = \frac{M_1 M_2}{M_1 + M_2} . \tag{A.13}$$

Equation (A.10) shows that the velocity of the centre of mass of the system remains constant, while equation (A.11) shows that the motion of one particle relative to the other is essentially that of a single particle of mass equal to the reduced mass, acted on by a force equal to the mutual interaction between the two particles.

Inverting equations (A.9), we have

$$r_1 = R + \frac{M_2}{M_1 + M_2} r \tag{A.14a}$$

$$r_2 = R - \frac{M_1}{M_1 + M_2} r . \tag{A.14b}$$

Then with $v_i = \dot{r}_i$, etc, the kinetic energies (A.2) can be expressed as

$$T = \tfrac{1}{2}MV^2 + \tfrac{1}{2}\mu v^2 \tag{A.15a}$$

$$T' = \tfrac{1}{2}M'V'^2 + \tfrac{1}{2}\mu' v'^2 \tag{A.15b}$$

respectively, where $V = \dot{R}$ is the velocity of the centre of mass before impact (constant), $v = \dot{r}$ is the relative velocity of the two particles before impact (variable), while V' and v' are the corresponding quantities after impact. Also

$$M = M_a + M_A \tag{A.16a}$$

$$M' = M_b + M_B \tag{A.16b}$$

$$\mu = \frac{M_a M_A}{M_a + M_A} \tag{A.17a}$$

$$\mu' = \frac{M_b M_B}{M_b + M_B} . \tag{A.17b}$$

Since we are assuming that the two-body system is non-relativistic both before and after impact, then Q, as defined by equation (A.5), will be so small that as far as kinematics are concerned we can set

$$M = M' . \tag{A.18}$$

Overall momentum conservation then leads to

$$V = V' \tag{A.19}$$

and

$$\tfrac{1}{2}MV^2 = \tfrac{1}{2}M'V'^2 . \tag{A.20}$$

If we transform the asymptotic velocities in the same way we find for equations (A.3)

$$E = \tfrac{1}{2}MV^2 + \tfrac{1}{2}\mu u^2 \tag{A.21a}$$

$$E' = \tfrac{1}{2}M'V'^2 + \tfrac{1}{2}\mu' u'^2 \tag{A.21b}$$

so that equation (A.4) becomes

$$\tfrac{1}{2}\mu' u'^2 = \tfrac{1}{2}\mu u^2 + Q . \tag{A.22}$$

Now for the reaction to proceed at all the asymptotic relative velocity u' of the product particles must exceed zero, since otherwise they would not separate. Thus we must have

$$\tfrac{1}{2}\mu u^2 + Q > 0 \tag{A.23}$$

so that in the case of an endothermic reaction ($Q < 0$) the threshold condition

$$\tfrac{1}{2}\mu u^2 > |Q| \tag{A.24}$$

has to be satisfied. More generally, we see that it is only the second term in equation (A.21a), $\tfrac{1}{2}\mu u^2$, that is available for the reaction, the first term being essentially wasted in motion of the centre of mass. We write

$$\tfrac{1}{2}\mu u^2 = E_{\text{CM}} \tag{A.25}$$

referring to it loosely as the *centre of mass* (CM) energy, this being the kinetic energy measured in the frame in which the centre of mass is at rest.

In the case of a laboratory experiment, where the target nucleus is initially at rest, we have

$$\boldsymbol{u} = \boldsymbol{u}_{\text{a}} \tag{A.26}$$

from which it follows that the available CM energy is

$$\begin{aligned} E_{\text{CM}} &= \tfrac{1}{2}\mu u_{\text{a}}^2 \\ &= \frac{M_{\text{A}}}{M_{\text{a}} + M_{\text{A}}} E_{\text{lab}} \simeq \frac{A_{\text{A}}}{A_{\text{a}} + A_{\text{A}}} E_{\text{lab}} \end{aligned} \tag{A.27}$$

where we have introduced the mass numbers A of the different nuclei, and E_{lab} denotes the energy of the projectile measured in the laboratory frame of reference†. Thus the lighter the projectile compared with the target nucleus the less the degradation of energy in centre of mass motion. The threshold energy in the case of an endothermic reaction is, according to equation (A.24)

$$E^0_{\text{lab}} = \frac{A_{\text{a}} + A_{\text{A}}}{A_{\text{A}}} \,|\,Q\,|\,. \tag{A.28}$$

(It is left as an instructive exercise to determine the threshold photon energy for an endothermic photonuclear reaction—see Problem IV.2.)

More generally, we shall have from equation (A.22)

$$u' = \left(\frac{\mu u_{\text{a}}^2 + 2Q}{\mu'}\right)^{1/2}. \tag{A.29}$$

Also, from equation (A.9a) the constant velocity of the centre of mass is

$$\boldsymbol{V} = \frac{M_{\text{a}}}{M}\,\boldsymbol{u}_{\text{a}}\,. \tag{A.30}$$

We have seen in equation (A.15) that the total kinetic energy (A.2), which is the sum of two one-body terms, can be transformed into a sum of two different one-body terms, one of which corresponds to a body of mass M and is associated with the motion of the centre of mass, and the other of which corresponds to a body of mass μ and is associated with the relative motion. An exactly analogous transformation is possible for the total orbital angular momentum: equation (A.14) leads to

$$\begin{aligned} \boldsymbol{L} &\equiv M_1(\boldsymbol{r}_1 \times \boldsymbol{v}_1) + M_2(\boldsymbol{r}_2 \times \boldsymbol{v}_2) \\ &= M(\boldsymbol{R} \times \boldsymbol{V}) + \mu(\boldsymbol{r} \times \boldsymbol{v})\,. \end{aligned} \tag{A.31}$$

†E_{CM} is lower than E_{lab} by virtue of the projectile mass M_{a} having been replaced by the reduced mass μ.

Thus with no external force acting the first term will be constant, as will the second term if the force of interaction is central.

The foregoing discussion is entirely classical, whereas in reality a quantum-mechanical treatment is required for nuclear particles. However, all the principal results will remain valid. In particular the characteristic separation into centre of mass motion and relative motion will still hold in quantum mechanics: the two-body Schrödinger equation separates into two one-body equations, one of which, for a particle of mass M, relates to the motion of the centre of mass, and the other of which, for a particle of mass μ, relates to the relative motion. It follows that all the foregoing results concerning threshold energies and asymptotic velocities remain valid.

Any theory of the interaction between two particles will naturally be formulated in terms of the relative motion: classically, for example, one would solve equation (A.11) for $\boldsymbol{r}$, or $\boldsymbol{r}'$, the separation vector of the two particles after impact. Experimentally, however, one observes the asymptotic directions θ_b, θ_B of the product particles, i.e. the direction of $\boldsymbol{u}_\mathrm{b}$ and $\boldsymbol{u}_\mathrm{B}$ in a laboratory-fixed frame, and it is necessary to relate these to the direction θ of the relative-velocity vector $\boldsymbol{u}'$ with respect to the beam direction, $\boldsymbol{u}_\mathrm{a} = \boldsymbol{u}$.

From the transformation equations (A.14) we have

$$\boldsymbol{u}_\mathrm{b} = \boldsymbol{V} + \frac{M_\mathrm{B}}{M}\boldsymbol{u}' \tag{A.32a}$$

$$\boldsymbol{u}_\mathrm{B} = \boldsymbol{V} - \frac{M_\mathrm{b}}{M}\boldsymbol{u}' \,. \tag{A.32b}$$

Then referring to figure A.1(b) and using equation (A.30) for $\boldsymbol{V}$ we have

$$\tan\theta_\mathrm{b} = \frac{\sin\theta}{\cos\theta + \gamma_\mathrm{b}} \tag{A.33a}$$

$$\tan\theta_\mathrm{B} = \frac{\sin\theta}{\gamma_\mathrm{B} - \cos\theta} \tag{A.33b}$$

where

$$\gamma_\mathrm{b} = \frac{M_\mathrm{a}u_\mathrm{a}}{M_\mathrm{B}u'} = \left(\frac{M_\mathrm{a}M_\mathrm{b}}{M_\mathrm{A}M_\mathrm{B}}\frac{E_\mathrm{CM}}{E_\mathrm{CM}+Q}\right)^{1/2} \tag{A.34a}$$

and

$$\gamma_\mathrm{B} = \frac{M_\mathrm{a}u_\mathrm{a}}{M_\mathrm{b}u'} = \left(\frac{M_\mathrm{a}M_\mathrm{B}}{M_\mathrm{A}M_\mathrm{b}}\frac{E_\mathrm{CM}}{E_\mathrm{CM}+Q}\right)^{1/2} . \tag{A.34b}$$

With the line Bb in figure A.1(b) representing the vector $\boldsymbol{u}'$, it follows that the point O denotes the centre of mass. Thus the angle θ represents the angle of emergence of the particles b and B in a frame of reference in which the centre of mass is at rest, and is usually referred to as such, $\theta \equiv \theta_{\mathrm{CM}}$. However, it is clear that θ also represents the angle of emergence in every frame for which any point on the line bB is at rest.

In reality, particles will emerge in all directions, and the interesting quantity will be the differential cross section as a function of angle, defined in equation (19.22). Again, theoretical treatments will give us the cross section in the CM frame of reference, $\sigma(\theta)$, while experimentally one measures the cross section in the laboratory-fixed frame, $\sigma(\theta_{\mathrm{b}})$, or $\sigma(\theta_{\mathrm{B}})$. From equation (19.23) we have

$$\sigma(\theta_{\mathrm{b}}) \sin\theta_{\mathrm{b}}\, \mathrm{d}\theta_{\mathrm{b}} = \sigma(\theta) \sin\theta\, \mathrm{d}\theta \tag{A.35}$$

and from equation (A.33a)

$$\sigma(\theta_{\mathrm{b}}) = \frac{(1 + 2\gamma_{\mathrm{b}} \cos\theta + \gamma_{\mathrm{b}}^2)^{3/2}}{1 + \gamma_{\mathrm{b}} \cos\theta}\, \sigma(\theta) \tag{A.36}$$

with a similar expression for $\sigma(\theta_{\mathrm{B}})$.

APPENDIX B SPIN FUNCTION OF TWO FERMIONS

Consider any angular-momentum observable

$$J = (J_x, J_y, J_z)\,. \tag{B.1}$$

Then basic to the quantum-mechanical theory of angular momentum are the commutation relations

$$[\hat{J}_x, \hat{J}_y] = \mathrm{i}\hbar\hat{J}_z,\ [\hat{J}_y, \hat{J}_z] = \mathrm{i}\hbar\hat{J}_x,\ [\hat{J}_z, \hat{J}_x] = \mathrm{i}\hbar\hat{J}_y \tag{B.2}$$

$$[\hat{J}^2, \hat{J}_x] = [\hat{J}^2, \hat{J}_y] = [\hat{J}^2, \hat{J}_z] = 0 \tag{B.3}$$

when $\hat{q}$ denotes the operator corresponding to the observable q.

It follows that there exist states which are simultaneously eigenstates of $\hat{J}^2$ and just one component of $\hat{J}$, which we conventionally take as J_z. We have, in fact

$$\hat{J}^2\psi_J^{J_z} = J(J+1)\hbar^2\psi_J^{J_z} \tag{B.4}$$

$$\hat{J}_z\psi_J^{J_z} = J_z\hbar\psi_J^{J_z} \tag{B.5}$$

where J can take any integral or half-integral value, while J_z can take any of the $2J+1$ values $J, J-1, J-2, \ldots, -J+1, -J$.

Consider now the intrinsic spin of a single fermion

$$s = (s_x, s_y, s_z)\,. \tag{B.6}$$

Here we have $J \equiv s = \frac{1}{2}$, so that $s_z = \pm\frac{1}{2}$. We denote the eigenstates of $\hat{s}_z$ by α and β, thus

$$\hat{s}_z\alpha = \tfrac{1}{2}\hbar\alpha \tag{B.7a}$$

$$\hat{s}_z\beta = -\tfrac{1}{2}\hbar\beta \tag{B.7b}$$

$$s^2\alpha = \tfrac{3}{4}\hbar^2\alpha \tag{B.8a}$$

$$s^2\beta = \tfrac{3}{4}\hbar^2\beta\ . \tag{B.8b}$$

A convenient representation of these eigenstates is

$$\alpha = \begin{pmatrix} 1 \\ 0 \end{pmatrix} \tag{B.9a}$$

$$\beta = \begin{pmatrix} 0 \\ 1 \end{pmatrix} \tag{B.9b}$$

from which

$$\hat{s}_z = \tfrac{1}{2}\hbar\begin{pmatrix} 1 & 0 \\ 0 & -1 \end{pmatrix}\ . \tag{B.10a}$$

The commutation rules (B.2) then lead to

$$\hat{s}_x = \tfrac{1}{2}\hbar\begin{pmatrix} 0 & 1 \\ 1 & 0 \end{pmatrix} \tag{B.10b}$$

$$\hat{s}_y = \tfrac{1}{2}\hbar\begin{pmatrix} 0 & -\mathrm{i} \\ \mathrm{i} & 0 \end{pmatrix}\ . \tag{B.10c}$$

It follows at once that

$$\hat{s}^2 = \tfrac{3}{4}\hbar^2 \tag{B.11}$$

and equation (B.8) is satisfied

The following useful results can be established easily

$$\hat{s}_x\alpha = \tfrac{1}{2}\hbar\beta \tag{B.12a}$$

$$\hat{s}_x\beta = \tfrac{1}{2}\hbar\alpha \tag{B.12b}$$

$$\hat{s}_y\alpha = \tfrac{1}{2}\mathrm{i}\hbar\beta \tag{B.12c}$$

$$\hat{s}_y\beta = -\tfrac{1}{2}\mathrm{i}\hbar\alpha. \tag{B.12d}$$

Consider now the combined spin of two fermions

$$\boldsymbol{S} = \boldsymbol{s}_1 + \boldsymbol{s}_2\ . \tag{B.13}$$

For the spin eigenstates of the pairs of fermions we can write

$$\hat{S}^2\chi_S^{S_z}(1,2) = S(S+1)\hbar^2\chi_S^{S_z}(1,2) \tag{B.14a}$$

$$\hat{S}_z\chi_S^{S_z}(1,2) = S_z\hbar\chi_S^{S_z}(1,2) \tag{B.14b}$$

where the possible values of S are 0 or 1, while S_z takes the value 0 in the former case, and ± 1 or 0 in the latter.

With

$$S_z = s_{1z} + s_{2z} \tag{B.15}$$

and, using equation (B.11)

$$\begin{aligned}\hat{S}^2 &= (\hat{s}_1 + \hat{s}_2)^2 = \hat{s}_1^2 + \hat{s}_2^2 + 2\hat{s}_1 \cdot \hat{s}_2 \\ &= \tfrac{3}{2}\hbar^2 + 2(\hat{s}_{1x}\hat{s}_{2x} + \hat{s}_{1y}\hat{s}_{2y} + \hat{s}_{1z}\hat{s}_{2z})\end{aligned} \tag{B.16}$$

it is easy to show, using equations (B.7) and (B.12), that $\chi_S^{S_z}(1,2)$ takes the following forms, according to the eigenvalues S and S_z

$$\begin{array}{lll} S_z = 1 & \alpha_1\alpha_2 & \\ S_z = -1 & \beta_1\beta_2 & S = 1 \\ S_z = 0 & (\alpha_1\beta_2 + \alpha_2\beta_1)/\sqrt{2} & \\ S_z = 0 & (\alpha_1\beta_2 - \alpha_2\beta_1)/\sqrt{2} & S = 0\,. \end{array} \tag{B.17}$$

(The $1/\sqrt{2}$ factors assure normalisation.) The three $S = 1$ states, known as the *triplet* states, are seen to be symmetric under exchange of the two fermions, while the $S = 0$ state, the so-called *singlet* state, is antisymmetric.

APPENDIX C GROUND-STATE ENERGY OF A FERMI GAS

Consider a single particle inside a cubic box with rigid walls. We suppose that apart from the constraints imposed by the walls no forces are acting on the particle inside the box, so that its kinetic energy ε will be constant, while the potential energy can be set equal to zero. The wavefunction of the particle will then satisfy the Schrödinger equation

$$-\frac{\hbar^2}{2M}\nabla^2\Psi = \varepsilon\Psi \tag{C.1}$$

within the box, and will have to vanish on the walls, since these cannot be penetrated at all, being supposed to be rigid. Then if we take the origin of the coordinates at one corner of the box, and the three axes along the edges, the wavefunction of our particle will take the form

$$\Psi = \sqrt{\frac{8}{a^3}}\left(\sin\frac{n_x\pi}{a}x\right)\left(\sin\frac{n_y\pi}{a}y\right)\left(\sin\frac{n_z\pi}{a}z\right) \tag{C.2}$$

where a is the side of the cube, the factor $\sqrt{8/a^3}$ assures a correct normalisation, and n_x, n_y, n_z are integers defining the energy† according to

$$\frac{\hbar^2}{2M}\frac{\pi^2}{a^2}(n_x^2+n_y^2+n_z^2)=\varepsilon. \tag{C.3}$$

It is important to realise that none of these integers can be zero: if any of them were then the wavefunction (C.2) would vanish. This means that the minimum possible energy of a particle confined in a box is finite

$$\varepsilon_0=\frac{3\hbar^2}{2M}\frac{\pi^2}{a^2} \tag{C.4}$$

and not zero as it would be according to classical mechanics. In fact, the presence of the factor $1/a^2$ means that the smaller the box the greater this so-called zero-point energy, a result that is intimately related to the Heisenberg principle, as discussed in §13.

Suppose now that there are N such particles in the box. If there are no interactions between the particles each one will still have a wavefunction of the form (C.2), and energy of the form (C.3), so that the total energy of the system will be

$$E=\frac{\hbar^2}{2M}\frac{\pi^2}{a^2}\sum_{i=1}^{N}[(n_x^i)^2+(n_y^i)^2+(n_z^i)^2] \tag{C.5}$$

where the triad of numbers (n_x^i, n_y^i, n_z^i) characterises the wavefunction and energy of the ith particle.

This total energy will be a minimum if for all particles $n_x=n_y=n_z=1$. Since there is only one such state, this means that all particles would be in the same state, which is certainly possible for bosons, i.e. particles having integral spin, such as photons and pions. However, we are concerned here with fermions, i.e. particles having spin $\frac{1}{2}\hbar$, and, according to the Pauli exclusion principle, there can never be more than one of these in any given state. Bearing in mind this constraint we now calculate the ground-state energy, i.e. minimum possible energy of our system of non-interacting fermions; we call this system a *Fermi gas*.

In applying the Pauli principle, it is first necessary to define what is meant by 'state', the degree of occupancy of which cannot exceed unity. Each triad of *positive* integers (n_x, n_y, n_z) will define a spatially distinct state, but negative integers will give nothing new, since changing the sign of one of these just changes the overall sign of the wavefunction (C.2). However, the Pauli principle distinguishes between the two possible directions of the fermion's spin, so that altogether we are allowed *two* fermions per set of positive integers (n_x, n_y, n_z).

† We are assuming here the non-relativistic relation between energy and momentum; this is always valid for the nucleons in a nucleus.

Let us suppose now that we put our N fermions into the box one at a time, always trying to keep the energy as low as possible. Just two can be accommodated in the state of lowest energy (1, 1, 1), so that if there are more than two particles we shall have to start filling other triads, which of necessity will have higher energy. We see at once a major consequence of the Pauli principle: it raises the lowest possible mean energy per particle in a fermion gas above the value it has in a boson gas of the same density†.

After (1, 1, 1), the lowest available states will be (2, 1, 1), (1, 2, 1) and (1, 1, 2), which are, of course, degenerate, and can hold six fermions in all. Clearly, the degree of degeneracy will increase with energy, so to be able to handle arbitrarily large numbers N of particles we calculate the number $\mathrm{d}\nu$ of distinct triads corresponding to the energy interval $(\varepsilon, \varepsilon + \mathrm{d}\varepsilon)$. To this end we represent the possible triads (n_x, n_y, n_z) by points in a three-dimensional number space: there will be one such point per unit volume. Points lying on a spherical surface of radius $\rho = (n_x^2 + n_y^2 + n_z^2)^{1/2}$ will, according to (C.3), all correspond to states of energy

$$\varepsilon = \frac{\hbar^2}{2M}\frac{\pi^2}{a^2}\rho^2 \tag{C.6}$$

while points lying within the spherical shell of radii ρ and $\rho + \mathrm{d}\rho$ will correspond to states of energy lying between ε and $\varepsilon + \mathrm{d}\varepsilon$, where

$$\mathrm{d}\varepsilon = \frac{\hbar^2}{M}\frac{\pi^2}{a^2}\rho\,\mathrm{d}\rho. \tag{C.7}$$

Now the number $\mathrm{d}\nu$ of such points in the shell is just equal to the volume of the shell, but only one octant of this can be considered, since we must restrict ourselves to positive values of n_x, n_y and n_z (see figure C.1, which, for simplicity, relates to a two-dimensional version of this calculation). Thus

$$\begin{aligned}\mathrm{d}\nu &= \tfrac{1}{2}\pi\rho^2\,\mathrm{d}\rho \\ &= \frac{V}{\pi^2\sqrt{2}}\left(\frac{M}{\hbar^2}\right)^{3/2}\varepsilon^{1/2}\,\mathrm{d}\varepsilon\end{aligned} \tag{C.8}$$

where $V = a^3$ is simply the volume of the box. Then with two particles per triad being allowed we see that the number $\mathrm{d}n$ of particles that can be accommodated in energy interval $\mathrm{d}\varepsilon$ is

$$\mathrm{d}n = \frac{\sqrt{2}}{\pi^2}\left(\frac{M}{\hbar^2}\right)^{3/2}\varepsilon^{1/2}\,\mathrm{d}\varepsilon \tag{C.9}$$

per unit volume.

† This is the second quantum effect tending to raise the minimum possible energy of a non-interacting gas above its classical value of zero, the other being the Heisenberg effect. However, the latter is operative even for a single particle, and also applies equally well to fermions and bosons.

The ground-state configuration of our Fermi gas will consist of all states up to a certain energy, known as the *Fermi energy*, ε_F, being filled, and all states above being empty. The value of ε_F will be determined through (C.9) by the number of particles per unit volume, $n = N/V$, thus

$$n = \frac{\sqrt{2}}{\pi^2}\left(\frac{M}{\hbar^2}\right)^{3/2}\int_0^{\varepsilon_F}\varepsilon^{1/2}\,d\varepsilon \tag{C.10}$$

from which

$$\varepsilon_F = \frac{\hbar^2}{2M}(3\pi^2 n)^{2/3}. \tag{C.11}$$

The total energy per unit volume will then be

$$E_V = \int_0^{\varepsilon_F}\varepsilon\,dn = \frac{\sqrt{2}}{\pi^2}\left(\frac{M}{\hbar^2}\right)^{3/2}\int_0^{\varepsilon_F}\varepsilon^{3/2}\,d\varepsilon$$

$$= \frac{1}{5\pi^2}\left(\frac{2M}{\hbar^2}\right)^{3/2}\varepsilon_F^{5/2} = \frac{3\hbar^2}{10M}(3\pi^2)^{2/3}n^{5/3}. \tag{C.12}$$

This gives for the average energy per particle

$$\bar{\varepsilon} = \frac{E_V}{n} = \frac{3}{5}\varepsilon_F. \tag{C.13}$$

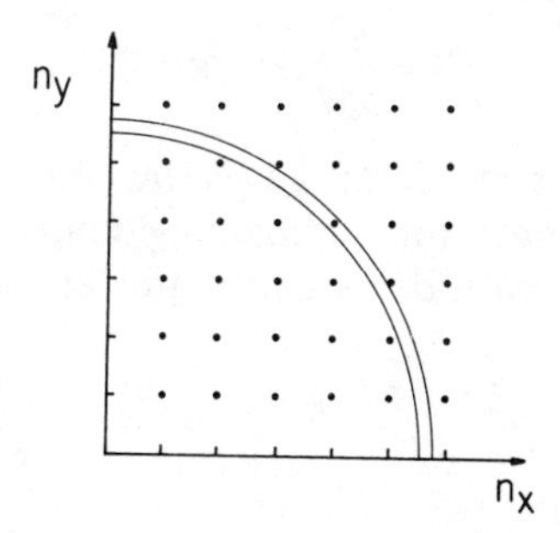

Figure C.1 Counting of permitted states (two-dimensional representation).

Higher-energy configurations of the Fermi gas can, of course, be realised simply by exciting particles to states above the Fermi energy ε_F. However, the system may be assumed to be effectively in the ground-state configuration provided it is in thermal equilibrium and the temperature is so low that

$$kT \ll \varepsilon_F. \tag{C.14}$$

Then the mean energy per particle will indeed be given by (C.13), and in particular will remain finite even as the temperature approaches absolute zero. This is in marked contrast to the predictions of classical, i.e. Maxwell–Boltzmann, statistics, according to which the energy per particle should vanish as $T \to 0$, being given by $\frac{3}{2}kT$. The extra energy is, as we have seen, simply a consequence of the Pauli principle.

Thus (C.14) is essentially the condition for specifically quantum-mechanical effects to become significant—we say that the gas becomes *degenerate*. At a given temperature, a gas is more likely to be degenerate the larger ε_F, i.e. the greater the density, according to (C.11). This is the origin of our statement in §13 that quantum-mechanical effects become more important with increasing density.

Equation of state of degenerate gas. Another significant difference between a degenerate gas and one obeying classical statistics concerns its pressure, P. On purely kinematic grounds we have in both cases, provided the particle velocities are non-relativistic

$$P = \tfrac{2}{3} E_V. \tag{C.15}$$

For a classical Maxwell–Boltzmann gas, where $E_V = \frac{3}{2}nkT$, this gives the familiar equation of state

$$P = nkT \tag{C.16}$$

while for a degenerate gas (C.12) gives

$$P = \frac{\hbar^2}{5M}(3\pi^2)^{2/3} n^{5/3}. \tag{C.17}$$

The pressure in a degenerate Fermi gas is thus remarkable firstly for being independent of the temperature, remaining finite even when $T \to 0$, and secondly for increasing with density at a greater than linear rate.

APPENDIX D GRAVITATIONAL CONTRACTION AND EVOLUTION OF STARS

Hydrostatic equilibrium. All stars must tend to contract under the gravitational attraction between their different parts. On the other hand, most stars appear not to be contracting, so presumably a pressure gradient is set up which balances the gravitational attraction.

To show that such a hydrostatic equilibrium is possible let us suppose that the star is spherical and does not rotate. Referring to figure D.1, consider a cylindrical element of stellar matter, base area ds and length dr, situated at distance r from the centre of the star with its axis lying along a radius vector. With P and ρ denoting pressure and density, respectively, an equilibrium will exist if

$$-\frac{\mathrm{d}P}{\mathrm{d}r}\,\mathrm{d}r\,\mathrm{d}s = G\frac{\mathcal{M}(r)\,\rho\,\mathrm{d}s\,\mathrm{d}r}{r^2}$$

i.e. if there is a pressure gradient at each point given by the so-called *equation of hydrostatic equilibrium*

$$\frac{\mathrm{d}P}{\mathrm{d}r} = -\rho(r)\frac{G\mathcal{M}(r)}{r^2} \tag{D.1}$$

where G is the gravitational constant and $\mathcal{M}(r)$ is the mass of material within the sphere of radius r

$$\mathcal{M}(r) = 4\pi\int_0^r \rho(r')r'^2\,\mathrm{d}r' \tag{D.2}$$

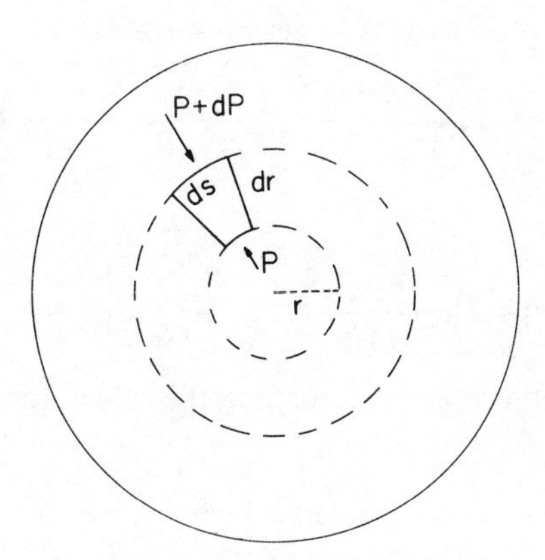

Figure D.1 Hydrostatic equilibrium within a star.

Before solving (D.1) we ask to what extent it remains valid even when we do not have true hydrostatic equilibrium. If a resultant force acts on the cylinder it will be accelerated, and in place of (D.1) we have

$$\frac{\mathrm{d}P}{\mathrm{d}r} = -\rho(r)\left(\frac{G\mathcal{M}(r)}{r^2} + \ddot{r}\right). \tag{D.3}$$

Now the first term here on the right-hand side is just the local gravitational acceleration, so that the second term will be negligible unless the contraction is so catastrophic as to resemble a free-fall collapse. Thus it is still possible for (D.1) to hold during a contraction; we say in such cases that the contraction is *quasistatic*.

To solve equation (D.1) let us make the simplifying assumption that the density of the star decreases linearly as we move away from the centre and vanishes on the surface

$$\rho(r) = \rho_0\left(1 - \frac{r}{R}\right) \tag{D.4}$$

where $\rho_0 \equiv \rho(0)$, the central density. Then from (D.2)

$$\mathcal{M}(r) = 4\pi\rho_0 \int_0^r \left(1 - \frac{r'}{R}\right) r'^2 \, dr'$$

$$= \frac{4\pi}{3} \rho_0 r^3 \left(1 - \frac{3}{4}\frac{r}{R}\right). \tag{D.5}$$

For $r = R$, the radius, this just gives the total mass $\mathcal{M}$ so that the central density is

$$\rho_0 = \frac{3}{\pi}\frac{\mathcal{M}}{R^3} = 4\bar{\rho} \tag{D.6}$$

where $\bar{\rho}$ is the mean density. We get now for (D.1)

$$\frac{dP}{dr} = -\frac{36G\mathcal{M}^2}{\pi R^6}\left(\frac{r}{3} - \frac{7}{12R}r^2 + \frac{r^3}{4R^2}\right)$$

from which

$$P(r) = P_0 - \frac{6G\mathcal{M}^2}{\pi R^6} r^2 \left(1 - \frac{7}{6}\frac{r}{R} + + \frac{3r^2}{8R^2}\right). \tag{D.7}$$

Assuming zero pressure on the surface this gives for the pressure at the centre

$$P_0 = \frac{5}{4\pi}\frac{G\mathcal{M}^2}{R^4}. \tag{D.8}$$

In the case of the sun, for which the mass is $\mathcal{M}_\odot = 1.98 \times 10^{30}$ kg, and the radius $R_\odot = 6.96 \times 10^5$ km, (D.6) and (D.8) give respectively: $\rho_0 \simeq 6 \times 10^3 \text{ kg m}^{-3}$ and $P_0 = 5 \times 10^9$ atmospheres (1 atmosphere $\simeq 10^5 \text{ N m}^{-2}$). Both of these results are an order of magnitude too small, the known values being about 10^5 kg m^{-3} and 10^{11} atmospheres, so obviously the assumption of constant density gradient is far from exact. On the other hand, it will not be totally misleading as a qualitative guide, and, for want of anything better, we shall retain it.

However, for the moment we are not concerned so much with precise numerical quantities as with simply making the point that solutions to (D.1) do exist, so that it should indeed be possible for a star to be in a state of hydrostatic equilibrium, with the tendency to contraction successfully resisted by the internal pressure. Nevertheless, we shall now see that unless a star has some internal source of energy, such as nuclear, to supply the energy that it is obviously radiating out into space, then the star must inevitably succumb to the tendency to contract.

Suppose that the star does not contract. Then the energy that is radiated away must be drawn from the internal thermal energy of the star, i.e. the star cools (we are supposing that it is behaving as a classical, non-degenerate gas). But then in cooling the star will lose internal pressure, so that the equation of hydrostatic equilibrium (D.1) will no longer be satisfied, in which case contraction will ensue.

Furthermore, once the star is contracting, we have a source for the energy that is being radiated away: it comes simply from the diminishing gravitational potential energy, Ω, which is given in general by

$$\Omega = -G\int_0^R \frac{\mathcal{M}(r)\mathrm{d}\mathcal{M}}{r}. \tag{D.9}$$

If we assume the model of linear density variation (D.4), this becomes†

$$\Omega = -\frac{26}{35}\frac{G\mathcal{M}^2}{R}. \tag{D.10}$$

Of course, now that we have a source for the energy that is being radiated away, there is no reason why the star should cool down. In fact, we now see that *as the star contracts its temperature rises*. Let us assume that the material of the star behaves as a classical ideal gas, so that the equation of state is

$$P = \frac{R_{\mathrm{gas}}}{\mu}\rho T \tag{D.11}$$

where μ is the mean atomic weight per particle, and $R_{\mathrm{gas}} \equiv N_{\mathrm{A}}k = 8.31 \times 10^3\ \mathrm{J\,kmol^{-1}\,K^{-1}}$ is the gas constant. Combining now (D.6), (D.8) and (D.11), we get for the temperature at the centre of a star‡

$$T_0 = \frac{5}{12}\frac{G}{R_{\mathrm{gas}}}\mu\frac{\mathcal{M}}{R} \tag{D.12}$$

which increases as the radius decreases, i.e. as the star contracts (note that we are supposing that the contraction is quasistatic so that (D.1), and its solution, remain valid).

† If on the other hand we assume constant density out to the surface of the star, we shall have $\Omega = -(3/5)G\mathcal{M}^2/R$, in analogy with the electrostatics result (14.7); the minus sign is simply a consequence of the gravitational force being attractive.

‡ Considering the specific case of the sun, let us suppose that this consists entirely of completely ionised hydrogen. This means that there will be approximately two free particles per AMU, so that $\mu \simeq 1/2$. Then for the central temperature we find $T_0 = 5 \times 10^6$ K, which is correct to within an order of magnitude.

This heating up that accompanies the contraction of the star means that only a part of the released gravitational energy is being radiated away, the remainder going into increasing the star's internal thermal energy. In fact, there is a very general result, the *virial theorem*, which asserts that the gravitational energy must be partitioned equally between these two modes, i.e. half is radiated away, and half is converted into thermal energy.

Proof of the virial theorem†. Multiplying both sides of (D.1) by $V(r)\,\mathrm{d}r = (4\pi/3)r^3\,\mathrm{d}r$ gives

$$V(r)\,\mathrm{d}P = -\frac{1}{3}\frac{G\mathcal{M}(r)}{r}\,\mathrm{d}\mathcal{M} = \frac{1}{3}\,\mathrm{d}\Omega \qquad \text{(D.13)}$$

where $\mathrm{d}\Omega$ is the gravitational potential energy of the shell of mass $\mathrm{d}\mathcal{M}$ and radius r, the pressure difference across which is $\mathrm{d}P$. Then the total gravitational potential energy of the star is

$$\Omega = 3\int_0^R V(r)\,\mathrm{d}P = 3(V(r)P(r))_0^R - 3\int_0^R P(r)\,\mathrm{d}V.$$

This gives, assuming that the pressure on the surface vanishes,

$$\Omega = -3\int_0^R P(r)\,\mathrm{d}V. \qquad \text{(D.14)}$$

Now if the gas is monatomic, so that the internal energy is entirely translational, we shall have

$$P = \tfrac{2}{3}u \qquad \text{(D.15)}$$

where u is the density of internal energy (this result remains valid if the gas is ionised, but requires that the electrons be non-relativistic). Combining (D.14) and (D.15) we have for the total internal energy of the star

$$U = -\tfrac{1}{2}\Omega \qquad \text{(D.16)}$$

so that half of any loss of gravitational energy must indeed be converted into internal energy:

$$\Delta U = -\tfrac{1}{2}\Delta\Omega. \qquad \text{(D.17)}$$

Note that this result does not depend on the adoption of the model of constant density gradient, defined by equation (D.4), that we were using before.

Nuclear energy source. Having established gravitational contraction as a viable source of stellar energy, the question arises as to whether it is the only one. If it were, then from (D.10) and (D.17) we would have for the

† This subsection can be omitted without loss of continuity.

luminosity of the star, i.e. the total rate of emission of energy, assuming that the star followed the density profile (D.4)

$$L = -\frac{1}{2}\frac{\mathrm{d}\Omega}{\mathrm{d}t} = -\frac{13}{35}\frac{G\mathcal{M}^2}{R}\frac{1}{R}\frac{\mathrm{d}R}{\mathrm{d}t}. \tag{D.18}$$

Considering the specific case of the sun, we find that the fractional rate of contraction that would be required to account for the present luminosity of $L_\odot = 3.9 \times 10^{26}\ \mathrm{J\,s^{-1}}$ is $(\mathrm{d}R/R\ \mathrm{d}t) \simeq 10^{-7}$ per year (the values of $\mathcal{M}_\odot$ and $R_\odot$ are given above). This is too slow a contraction to be observable, so that there is no real incompatibility between gravitational contraction and the apparent hydrostatic equilibrium that prevails.

However, if we integrate (D.18) over the history of the sun, assuming infinite radius at the time it began to condense out of the diffuse galactic gas, then we find that the sun could have been shining with its present luminosity for no more than 10^7 years, if gravity were the only source of energy. But there is overwhelming geological evidence that the sun's luminosity has been roughly constant† for at least 10^9 years, so some alternative energy source must be operating. We require in fact at least 10^{43} J generated over the sun's history up to the present time, and since there are about 10^{57} nucleons in the sun this makes $10^{-14}\ \mathrm{J} \simeq 50$ keV per H atom. Now chemical reactions typically involve energy changes of the order of an eV per atom, so these are obviously quite inadequate. The only other possibility is nuclear reactions and we see at once that these are sufficient, since the conversion of H to ^{4}He yields about 7 MeV per nucleon (maximum energy release corresponds to the conversion of H to ^{56}Fe, but going all the way from ^{4}He to ^{56}Fe adds only another 10% or so).

Now although gravitational contraction can supply only a minor fraction of the total energy output of a star such as the sun (we know, incidentally, that the sun is a quite typical star), the attendant temperature rise plays a vital role in triggering the nuclear reactions that do account for the bulk of the star's energy output. This is because for two colliding nuclei to react, their relative kinetic energy must be sufficiently high for there to be an appreciable penetration of the Coulomb barrier between the two nuclei, and in the manifest absence of any accelerator or other artificial device, the only way in which the nuclei can have this energy is by virtue of their thermal motion (see Chapters IV and VI and Appendix E).

What happens, then, is that a newly-born star, consisting mainly of hydrogen, contracts gravitationally and its temperature rises until the nuclear reactions that transmute hydrogen into helium can begin. The rate of these reactions increases very rapidly with the rising temperature, as discussed in Appendix E, until the rate of generation of nuclear energy is

† The ice ages constitute relatively minor perturbations.

equal to the luminosity† of the star, i.e. the rate at which it is radiating out into space. At this point the contraction must stop, since with all the radiated energy being drawn from the nuclear source none of the gravitational energy released by contraction could be radiated away. Rather it would all be converted into internal thermal energy, which would thus increase faster than permitted by the virial theorem. Hydrostatic equilibrium would then be violated and the star would reflate, the temperature, and hence pressure, being excessive.

This 'brake' on the gravitational contraction is very effective, since if there is any overshoot the consequent rise in temperature will increase the rate of nuclear energy generation, thereby aggravating the imbalance. Of course, when contraction ceases the temperature will stop rising; it is somewhat paradoxical that this is brought about by the onset of a new energy source.

After an initial phase of gravitational contraction, then, a star will settle down into a state of true hydrostatic equilibrium, remaining at constant temperature, and shining entirely by virtue of the nuclear energy released in the conversion of hydrogen into helium: the so-called 'main-sequence' stars are in this situation. As for the helium that is formed, the Coulomb barrier for this is four times higher than for hydrogen, so that it will not burn at the equilibrium temperature for hydrogen burning. Thus, after a significant amount of hydrogen has been burned the rate of nuclear energy generation will fall, whereupon gravitational contraction will resume and continue until the attendant temperature rise permits helium burning on a sufficient scale.

Thus, the overall history of the star is one of gravitational contraction punctuated by phases of nuclear burning, the fuel of one such phase being the ashes of the previous one. This could continue until the star had been transformed into ^{56}Fe, after which it would be impossible to generate any more energy through nuclear reactions, but before this stage was reached the process might have been brought to a halt by the circumstances discussed below.

White dwarfs. Suppose that our star has contracted so far that the number density n of the electron gas has increased to the point that it is essentially degenerate. According to equations (C.11) and (C.14) this will happen when

$$n \equiv \frac{N_A \rho}{\mu_e} \gg \frac{(2m_e kT)^{3/2}}{3\pi^2 \hbar^3} \qquad \text{(D.19)}$$

† Of course, the luminosity is not a fixed constant of the star but rather is a function of its composition and in particular of its temperature distribution, so that it will change as the star evolves. We have in fact a very complicated coupled situation—see the book by Clayton listed in the bibliography for Chapter VI.

where we have expressed n in terms of the mass density ρ, N_A being Avogadro's number, and μ_e is the mean atomic weight per electron. This latter quantity is given by Z/A for a completely ionised gas, which we assume to be the case here. It will thus be equal to 1 for hydrogen, and about 2 for heavier nuclei ($Z/A \simeq 1/2$) such as those found in a highly evolved star. Finally, since m_e denotes the electron mass, we see at once that the electron gas will be degenerate at much lower densities than the gas of nuclei, although the nuclei themselves are, of course, degenerate assemblies of nucleons.

Actually, not only does the density increase during contraction but so also does the temperature. However, we see from (D.12) that the latter only increases as $1/R$, while the density increases as $1/R^3$, so that contraction must lead sooner or later to the onset of degeneracy. Combining (D.19) with (D.12) and (D.6) shows that degeneracy at the centre of the star will begin when the temperature there has *risen* to the order of

$$T_{od} = \frac{2}{R_{gas}} \frac{m_e}{\hbar^2} \frac{1}{N_A^{5/3}} \left(\frac{\mu_e}{9\pi}\right)^{2/3} \left(\frac{5\mu G}{12}\right)^2 \mathcal{M}^{4/3}$$
$$\sim 6\left(\frac{\mathcal{M}}{\mathcal{M}_\odot}\right)^{4/3} \times 10^8 \text{ K} \qquad \text{(D.20)}$$

where for μ, the mean atomic weight per particle, we have again taken the value 2, as appropriate for heavy nuclei (the electrons are far more numerous than the nuclei).

Let us now consider the pressure at the centre of a star when the electrons there have become degenerate. This will be determined, as always, by the mass and the radius of the star through (D.8), if the star is in hydrostatic equilibrium (we are still using our model of constant density gradient). However, the pressure of a degenerate gas† is determined uniquely by its density, since the degree of freedom associated with the temperature is now missing. Since the density at the centre of the star is, like the pressure, also determined by the mass and the radius of the star, this time through (D.6), it follows that when the electrons in a star have become completely degenerate only one value of the radius is possible for a given mass. Using the equation of state (C.17), and combining with (D.6), (D.8) and the first member of (D.19), we find for this unique radius

$$R_0 = \frac{\hbar^2}{m_e} (3\pi^2)^{2/3} \left(\frac{3N_A}{\pi\mu_e}\right)^{5/3} \frac{4\pi}{25G} \frac{1}{\mathcal{M}^{1/3}}$$
$$\simeq 0.013 \times \left(\frac{\mathcal{M}_\odot}{\mathcal{M}}\right)^{1/3} R_\odot \qquad \text{(D.21)}$$

where again we have taken $\mu_e \simeq 2$.

† The pressure exerted by the non-degenerate gas of nuclei will be negligible.

It would seem, then, that this is a limiting radius at which our star will stabilise: when it has contracted this far it can contract no further. At the same time, the temperature will not be able to rise any further either, so that the star will be unable to move on to the next phase of nuclear burning. The star ceases therefore to evolve and will start to cool down with the radius remaining at the constant value R_0†. Actually, we see from (D.20) that the heavier the star the hotter it will become before degeneracy sets in, so that the evolution of heavier stars proceeds further before being halted by degeneracy, i.e. they go through more phases of nuclear burning, approaching more closely the limit of ^{56}Fe, beyond which no further nuclear energy generation is possible.

Such an object will be considerably smaller than normal stars: if the mass is the same as that of the sun we see from (D.21) that its radius will be about a hundred times smaller, so that it will be about the same size as the earth, while the density will be of the order of tonnes per cm^3. At the same time the temperature at the centre of such a star, as given by (D.20), will be very much higher than in a less evolved star such as the sun. Being, then, both much smaller and much hotter than normal stars these objects, if they exist, should be easily recognised, and in fact they are identified with the familiar class of stars known as *white dwarfs*. Of course, as these stars cool down they will fade from view, doing so, because of their smallness, at relatively high temperatures.

We notice from (D.21) that the heavier the star the smaller the limiting radius R_0, i.e. the greater the electron density n and hence the greater the mean energy of the electrons. Now all the above discussion of white dwarfs supposes that the electrons are non-relativistic but it is clear that this will become less and less valid with increasing stellar mass.

It is fairly easy to understand qualitatively what the consequences of

† The virial theorem (D.16) is *always* valid, so that once the nuclear energy supply is shut off, these degenerate stars, like any other, can only shine by contracting. Thus the approach to the stable radius R_0 must be asymptotic: contraction only stops when the temperature of the star has fallen to zero. However, long before this happens the radius may be immeasurably close to R_0, a very slow rate of contraction being sufficient, in view of the small radius. Since the internal energy must, as always, *increase* during the contraction, we might ask how such a star can be 'cooling off'. It should be realised that an increase in internal energy implies a temperature increase only in the case of a non-degenerate gas. Compressing a degenerate gas leads to an increase in the internal energy even though the temperature remains effectively constant at zero. In the present case, it may be shown that the entire loss of gravitational energy goes to increasing the electron internal energy in the contracting star without altering its essentially zero-temperature configuration. Thus the energy radiated by the star must be drawn entirely from the gas of nuclei, and since this is non-degenerate it will indeed be cooling down. (I thank M Tassoul and G Fontaine for discussions on this point.)

relativistic effects will be for our degenerate star. For a non-relativistic degenerate gas the mean speed of the particles varies as $\varepsilon_F^{1/2}$ and thus increases with the density according to $n^{1/3}$. But because of the limiting value c this mean speed will increase less rapidly with n once it becomes comparable to c and in the limit of an extremely relativistic gas it will be independent of n. Thus we may expect the pressure to increase more slowly with density than in the case of a non-relativistic degenerate gas, and in fact it is easily shown that instead of the $n^{5/3}$ dependence implied by (C.17) we shall tend to have rather an $n^{4/3}$ dependence.

Consider now a star that is heavy enough for its degenerate electrons to have become relativistic before it has contracted down to the limiting radius given by the *non-relativistic* equation (D.21). Thereafter the pressure of the degenerate electrons will rise more slowly than had been supposed and will never be sufficient to withstand the gravitational attraction, so that instead of forming a stable white dwarf our star will collapse under its own weight, as it were. This collapse will be interrupted from time to time by the onset of a new phase of nuclear burning leading to a sufficient rate of energy generation but once the core of the star has transmuted to ^{56}Fe all further reactions will be endothermic, and neither nuclear burning nor the electron degeneracy pressure will be able to stop the collapse.

The critical mass above which a star cannot attain stability as a white dwarf is known as the *Chandrasekhar limit*. To estimate this we assume to begin with that the electrons remain non-relativistic so that the stabilising radius is given by (D.21). The corresponding Fermi energy is then calculated, and we suppose that relativistic effects will be significant and collapse inevitable if ε_F is of the order of the electron's rest-mass energy, $m_e c^2$. We thus estimate the Chandrasekhar limit as

$$\mathcal{M}_c = 3(8\pi^2)^{1/4}\left(\frac{12\hbar c}{25G}\right)^{3/2}\left(\frac{N_A}{\mu_e}\right)^2 \simeq \frac{5.6}{\mu_e^2}\mathcal{M}_\odot \qquad \text{(D.22)}$$

which gives 1.4 $\mathcal{M}_\odot$†, assuming $\mu_e = 2$. Despite the crudeness of this order of magnitude estimate, it agrees very closely with the results of much more refined calculations.

Neutron stars. We now consider the consequences of this collapse of a star that has exhausted its nuclear fuel supply, having burned right through to ^{56}Fe, and which is too heavy to be stabilised as a white dwarf by the degeneracy pressure of the electrons. As the Fermi energy of the degenerate electrons in the ever-contracting star rises beyond 782 keV, they will begin to be captured through inverse β decay (see §16) by nuclear-bound protons,

† Many stars whose masses originally exceeded this limit are able to expel large quantities of matter at an advanced stage of their evolution without undergoing a supernova explosion, and thus are able to stabilise as white dwarfs after all.

transforming them into neutrons. Thus as the collapse continues, the nuclei become progressively richer in neutrons, forming nuclides that can never be observed under normal terrestrial conditions. Neutrons will begin to leak out when the neutron drip line is reached (see §17), and finally when the star has shrunk to the point that the overall density approaches that of ordinary nuclear matter, so that the nuclei in the star are touching each other for a significant fraction of the time, the nuclei will break up and form a gas that consists almost entirely of neutrons (there will be a small residual fraction of protons and an equal number of free electrons†, determined by the condition that the Fermi energy be equal to 782 keV).

Now just as the electron gas of the earlier phase of contraction inevitably became degenerate regardless of the temperature rise, so must this neutron gas (we ignore for simplicity the weak proton–electron admixture). Assuming for the moment that this gas remains non-relativistic, it must eventually stabilise at a radius given by an equation of the same form as (D.21), the only difference being that we shall have to replace the electron mass m_e by the neutron mass M, and the mean atomic weight per electron $\mu_e(\simeq 2)$ by the mean atomic weight per neutron $\mu_n(=1)$. Then

$$R_0 \simeq 2.2 \times 10^{-5} \left(\frac{\mathscr{M}_\odot}{\mathscr{M}}\right)^{1/3} R_\odot \qquad \text{(D.23)}$$

so that if the mass of our star is comparable with that of the sun the radius will be of the order of a few kilometres, the density being in the range 10^{17}–10^{18} kg m^{-3}. Such objects are simply the *neutron stars* discussed in §15, and their formation effectively brings to a halt the collapse of the defunct star.

Actually the termination of the collapse is both more complicated and more spectacular than we have suggested. The inner regions of the star, being more highly evolved, collapse faster than the outer regions, and eventually become detached. When the degeneracy pressure of the neutrons brings this collapse to a sudden halt, an enormous amount of energy is transmitted in one way or another to the envelope, which is blown off into space. This is a *supernova explosion*, and is the principal means by which the products of stellar nucleosynthesis are distributed throughout the galaxy (see §28). It is because of this mass loss that we may well find neutron stars with a mass lower than the Chandrasekhar limit for white-dwarf formation.

We now have to consider relativistic effects. If these are significant then it is quite certain that the neutron star will fail to stabilise, in exactly the same way as did the white dwarf. Still pursuing the analogy with the white dwarf, we see that this failure to stabilise will become more probable as the mass increases. Indeed, we can define a Chandrasekhar limit for formation

† At higher densities elementary particles that are normally unstable, such as muons, will make an appearance.

of stable neutron stars, noting that it will be given by an equation of the form of (D.22), the only difference being that we must replace $\mu_e(\simeq 2)$ by $\mu_n(=1)$, thus finding 5.6 $\mathcal{M}_\odot$.

Actually, the effects of both general relativity and the neutron–neutron interaction will modify this limit, and the real situation is somewhat confused. It seems that two solar masses is a more reasonable upper limit. Note, however, that a star heavier than this limiting value could still end up as a neutron star, provided it got rid of enough mass during the supernova explosion. But if such a heavy star does not lose enough mass it will not stabilise as a neutron star, and the contraction will continue unchecked towards formation of a so-called *black hole*, the discussion of which lies beyond the scope of the present book.

APPENDIX E THERMONUCLEAR REACTION RATES

In §19 we introduced the cross section as the parameter characterising the probability of a given reaction between nuclei a and A, and we showed how it could be extracted from laboratory measurements. We show here how the thermonuclear reaction rates can be expressed in terms of the appropriate cross section, even though the situation prevailing in stellar interiors (or in artificial thermonuclear devices) does not correspond at all to the usual laboratory set-up of well collimated beams bombarding thin, flat targets.

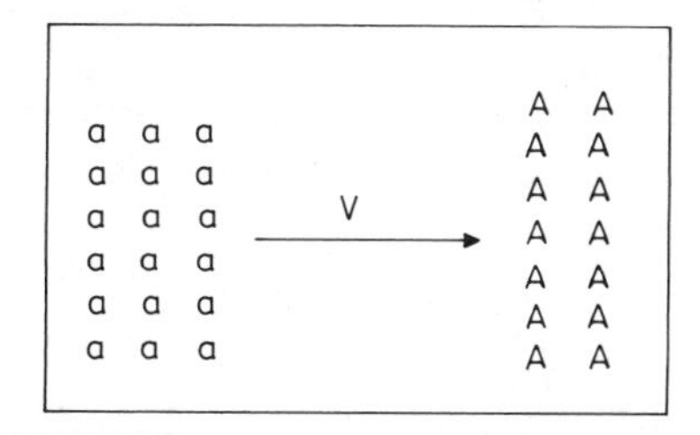

Figure E.1 Schema for calculation of thermonuclear reaction rates.

Referring to figure E.1, we suppose that all the nuclei of type a are initially separated from those of type A, although they are uniformly distributed within their respective regions, with the same densities that they have in the actual situation that interests us, n_a and n_A nuclei, respectively, per unit volume. We suppose furthermore that all the nuclei A are at rest, while all the nuclei a are projected towards the nuclei A with the same velocity v. The current density of these projectile nuclei a will then be $j = n_a v$, and it follows from equation (19.20) that the rate per unit volume of the reaction $a + A \rightarrow B + b$ will be

$$r_{aA,\,bB} = \sigma_{aA,\,bB}(v) v n_a n_A \tag{E.1}$$

where $\sigma_{aA,\,bB}$ denotes the integrated cross section for this reaction, and we have displayed its possible energy dependence. Henceforth for simplicity we shall drop the subscripts aA, bB.

Since in reality both types of nucleus are moving, the quantity v will become the relative speed between interacting pairs. As this will not take a unique value but rather will constitute a continuous spectrum, we shall denote by $\phi(v)\,dv$ the fraction of pairs (a,A) having relative speed between v and $v + dv$. Then we can write

$$r \equiv n_a n_A \langle \sigma v \rangle \tag{E.2}$$

where

$$\langle \sigma v \rangle = \int_0^\infty v \sigma(v) \phi(v)\,dv \tag{E.3}$$

in which

$$\int_0^\infty \phi(v)\,dv = 1. \tag{E.4}$$

If the reacting nuclei are identical then the number of reacting pairs is $\frac{1}{2} n_a^2$ and not n_a^2, so in general we should write equation (E.2) as

$$r = (1 + \delta_{aA})^{-1} n_a n_A \langle \sigma v \rangle. \tag{E.5}$$

Let us now calculate the spectrum of relative speeds $\phi(v)$ for the usual situation where the reacting nuclei a and A are in the form of perfect non-degenerate non-relativistic gases in thermal equilibrium. Thus the velocity distributions of the two nuclides will be Maxwellian

$$n_a(v_a)\,d^3 v_a = n_a \left(\frac{M_a}{2\pi kT}\right)^{3/2} \exp\left(-\frac{M_a v_a^2}{2kT}\right) d^3 v_a \tag{E.6}$$

and likewise for A. Then for the joint velocity distribution we have

$$n_a(v_a) n_A(v_A)\,d^3 v_a\,d^3 v_A =$$

$$n_a n_A \frac{(M_a M_A)^{3/2}}{(2\pi kT)^3} \exp\left(-\frac{M_a v_a^2 + M_A v_A^2}{2kT}\right) d^3 v_a\,d^3 v_A. \tag{E.7}$$

In order to be able to use equation (E.3) we must transform from the laboratory velocities $\mathbf{v}_a$ and $\mathbf{v}_A$ to the relative velocity

$$\mathbf{v} = \mathbf{v}_a - \mathbf{v}_A \tag{E.8a}$$

and the velocity of the centre of mass

$$V = \frac{M_A v_A + M_a v_a}{M_a + M_A}. \tag{E.8b}$$

Inverting these two equations we find

$$v_A = V - \frac{M_a}{M_a + M_A} v \tag{E.9a}$$

$$v_a = V + \frac{M_A}{M_a + M_A} v. \tag{E.9b}$$

Hence the total kinetic energy is

$$\begin{aligned} K &\equiv \tfrac{1}{2} M_A v_A^2 + \tfrac{1}{2} M_a v_a^2 \\ &= \tfrac{1}{2}(M_a + M_A) V^2 + \tfrac{1}{2} \mu v^2 \end{aligned} \tag{E.10}$$

where

$$\mu = \frac{M_a M_A}{M_a + M_A} \simeq M \frac{A_a A_A}{A_a + A_A} \tag{E.11}$$

is the reduced mass. From equation (E.9) we have for the Jacobian of the transformation

$$\begin{vmatrix} \dfrac{\partial v_{Ax}}{\partial V_x} & \dfrac{\partial v_{Ax}}{\partial v_x} \\ \dfrac{\partial v_{ax}}{\partial V_x} & \dfrac{\partial v_{ax}}{\partial v_x} \end{vmatrix} = 1 \tag{E.12}$$

and likewise for the y and z components. Thus we can rewrite equation (E.7) as

$$\begin{aligned} n_a(v_a) n_A(v_A)\, d^3v_a\, d^3v_A &\equiv n(v, V)\, d^3v\, d^3V \\ &= n_a n_A \frac{(M_a M_A)^{3/2}}{(2\pi kT)^3} \exp\left(-\frac{M_a + M_A}{2kT} V^2 - \frac{\mu}{2kT} v^2\right) d^3v\, d^3V. \end{aligned} \tag{E.13}$$

For the integral over V we have

$$\begin{aligned} \int \exp\left(-\frac{M_a + M_A}{2kT} V^2\right) d^3V &= 4\pi \int_0^\infty V^2 \exp\left(-\frac{M_a + M_A}{2kT} V^2\right) dV \\ &= \left(\frac{2\pi kT}{M_a + M_A}\right)^{3/2} \end{aligned} \tag{E.14}$$

so that

$$\begin{aligned} n(v)\, d^3v &\equiv d^3v \int n(v, V)\, d^3V \\ &= n_a n_A \left(\frac{\mu}{2\pi kT}\right)^{3/2} \exp\left(-\frac{\mu v^2}{2kT}\right) d^3v. \end{aligned} \tag{E.15}$$

Thus the relative velocity follows a Maxwellian distribution, based upon the reduced mass. Integrating over the directions of the relative velocity vector we have†

$$n(v)\,\mathrm{d}v = n_{\mathrm{a}}n_{\mathrm{A}}\sqrt{\frac{2}{\pi}}\left(\frac{\mu}{kT}\right)^{3/2} v^2 \exp\left(-\frac{\mu v^2}{2kT}\right)\mathrm{d}v \tag{E.16}$$

i.e.

$$\phi(v) = \sqrt{\frac{2}{\pi}}\left(\frac{\mu}{kT}\right)^{3/2} v^2 \exp\left(-\frac{\mu v^2}{2kT}\right) \tag{E.17}$$

so that

$$\begin{aligned}\langle\sigma v\rangle &\equiv \int_0^\infty \sigma(v)\phi(v)v\,\mathrm{d}v \\ &= \sqrt{\frac{2}{\pi}}\left(\frac{\mu}{kT}\right)^{3/2}\int_0^\infty \sigma(v)v^3 \exp\left(-\frac{\mu v^2}{2kT}\right)\mathrm{d}v \\ &= \left(\frac{8}{\pi\mu(kT)^3}\right)^{1/2}\int_0^\infty \sigma(E)E\exp\left(-\frac{E}{kT}\right)\mathrm{d}E\end{aligned} \tag{E.18}$$

where $E = \frac{1}{2}\mu v^2$ is the kinetic energy of the reacting nuclei in the centre of mass frame of reference.

To proceed further we have to make some assertion about how the cross section σ varies with energy. Excluding the case of neutron reactions, we shall suppose that the cross section is given by the Gamow expression (20.14)

$$\sigma(E) = \frac{S}{E}\exp\left(-\frac{b}{\sqrt{E}}\right) \tag{E.19}$$

where, using equation (20.13)

$$b = \gamma\left(\frac{A_{\mathrm{a}}A_{\mathrm{A}}}{A_{\mathrm{a}} + A_{\mathrm{A}}}\right)^{1/2} Z_{\mathrm{a}}Z_{\mathrm{A}} \tag{E.20}$$

with $\gamma = 31\sqrt{KeV}$. As for the S factor, it may be assumed to be roughly energy independent, provided (*a*) the bombarding energy E is well below the Coulomb barrier and (*b*) there are no resonances near the bombarding energy E. If these conditions are not both fulfilled then it will still be possible, of course, to parametrise the cross section in the form (E.19), but S will then vary very rapidly with energy, and the following analysis will be invalid.

Assuming, then, the constancy of S, we combine equations (E.18) and

† Note the distinction between $n(v)$ and $n(\mathbf{v})$: we have $n(v) = 4\pi v^2 n(\mathbf{v})$.

(E.19) to obtain

$$\langle \sigma v \rangle = \left(\frac{8}{\pi \mu (kT)^3} \right)^{1/2} S \int_0^\infty \exp\left(-\frac{E}{kT} - \frac{b}{E^{1/2}} \right) \mathrm{d}E. \qquad \text{(E.21)}$$

The integrand here consists of a product of two factors, one of which, the Maxwell–Boltzmann tail, decreases monotonically with energy, and the other of which, the Gamow curve, increases monotonically with energy (see figure E.2). The result is that the integrand is peaked (we speak of the *Gamow peak*) with a maximum at an energy given by

$$\begin{aligned} E_0 &= \left(\frac{bkT}{2} \right)^{2/3} \\ &= 1.22 \left(Z_a^2 Z_A^2 \frac{A_a A_A}{A_a + A_A} T_6^2 \right)^{1/3} \text{ keV} \end{aligned} \qquad \text{(E.22)}$$

where T_6 denotes the temperature in millions of degrees kelvin. Since the bulk of the reactions take place in the vicinity of this energy, E_0, it is known as the *effective thermal energy*.

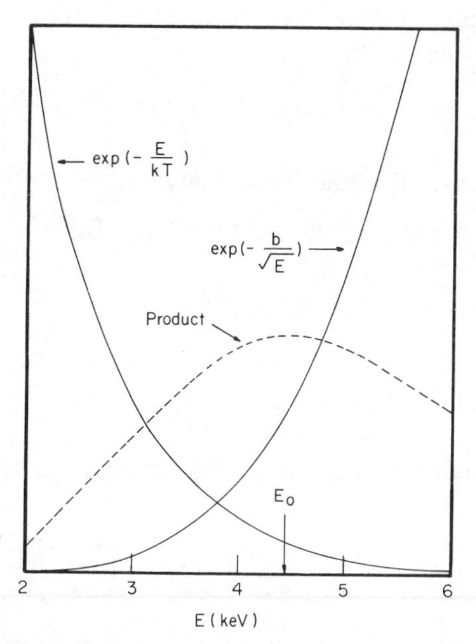

Figure E.2 Variation of $\exp(-E/kT)$ and $\exp(-b/\sqrt{E})$ with energy for a hydrogen-burning star; $T_6 = 10$ (see equation (E.21)).

The temperature of the central regions of a typical hydrogen-burning star is around 10^7 K, i.e. $T_6 \simeq 10$ so that $E_0 \simeq 5$ keV. Now from equation (18.3) we see that the height of the Coulomb barrier for two interacting hydrogen nuclei is about 0.6 MeV, so that the bulk of the reactions are indeed taking place well below the Coulomb barrier†, as we assumed in using the Gamow expression (E.19) for the cross section. On the other hand, with

$$kT = 0.086\, T_6 \text{ keV} \tag{E.23}$$

we see that in such stars kT is only about 1 keV, so that it is only nuclei far out in the tail of the Maxwell distribution that react significantly. The actual reaction rate is thus determined by the product of two very small factors, and is much lower than it would be if the star were hotter. Nevertheless, the rate of nuclear energy generation at these relatively low temperatures is still sufficient to provide all the star's output and thus to stabilise it.

Since E_0 varies only as $T^{2/3}$ it follows that the Maxwell peak will move towards E_0 as T increases, and in fact in a hydrogen-burning star the two will coincide at $T \simeq 10^9$ K. However, all hydrogen will have been burned long before such temperatures are reached, and the star will have moved on to burning much heavier nuclei, with considerably larger values of E_0 (§§28 and 29).

We make an approximate analytical evaluation of the integral in equation (E.21) by introducing the function

$$f(E) = \frac{E}{kT} + \frac{b}{E^{1/2}} \tag{E.24}$$

which is then expanded‡ about its maximum at E_0

$$f(E) \simeq f(E_0) + \tfrac{1}{2} f''(E_0)(E - E_0)^2 \tag{E.25}$$

where

$$f(E_0) = \frac{3E_0}{kT} \tag{E.26a}$$

$$f''(E_0) = \frac{3}{2E_0 kT}. \tag{E.26b}$$

Then the integrand in equation (E.21) becomes essentially a Gaussian centred at E_0

$$\int_0^\infty \exp(-f(E))\, \mathrm{d}E = \exp\left(-\frac{3E_0}{kT}\right) \int_0^\infty \exp\left[-\left(\frac{E - E_0}{\Delta/2}\right)^2\right] \mathrm{d}E \tag{E.27}$$

† For those reactions in hydrogen-burning stars that involve nuclei heavier than hydrogen (§27), the Coulomb barrier will be still higher.

‡ We are following this approximate analytical procedure in order to bring out the main physical features. For quantitative purposes it is always possible, of course, to evaluate the integral in equation (E.21) numerically, in which case one could take account also of the energy dependence of the S factor, which should be inside the integral.

where the full width is

$$\Delta = 4\sqrt{\frac{E_0 kT}{3}}$$

$$= 0.75\left(Z_a^2 Z_A^2 \frac{A_a A_A}{A_a + A_A} T_6^5\right)^{1/6} \text{ keV.} \qquad \text{(E.28)}$$

Again for the case of a typical hydrogen-burning star, $T_6 \simeq 10$, this gives $\Delta \simeq 5$ keV, which is comparable with E_0 itself, so that the Gamow peak is not particularly narrow (note, though, that Δ is the *full* width).

Replacing the lower limit of the integral in equation (E.27) by $-\infty$, equation (E.21) becomes

$$\langle \sigma v \rangle = \left(\frac{2}{\mu\,(kT)^3}\right)^{1/2} S(E_0)\,\Delta \exp\left(-\frac{3E_0}{kT}\right)$$

$$= 4\left(\frac{2E_0}{3\mu}\right)^{1/2} \frac{S(E_0)}{kT} \exp\left(-\frac{3E_0}{kT}\right) \qquad \text{(E.29)}$$

where, admitting the possibility of a *weak* energy dependence of S, we take its value at E_0. Since the quantity E_0 appearing in equation (E.29) is itself a function of the temperature, it should be eliminated if we are to display the temperature dependence of $\langle \sigma v \rangle$ clearly. This can be done most conveniently by introducing the dimensionless parameter

$$\tau = \frac{3E_0}{kT} = 42\left(Z_a^2 Z_A^2 \frac{A_a A_A}{A_a + A_A} \frac{1}{T_6}\right)^{1/3}. \qquad \text{(E.30)}$$

Equation (E.29) then becomes

$$\langle \sigma v \rangle = 7.2 \times 10^{-19} \frac{1}{Z_a Z_A} \frac{A_a + A_A}{A_a A_A} S(E_0)\tau^2 e^{-\tau} \text{ cm}^3\,\text{s}^{-1} \qquad \text{(E.31)}$$

where we have expressed S in units of barns keV.

Equation (E.31) shows that $\langle \sigma v \rangle$ has a maximum at $\tau = 2$, i.e. at $E_0 = 2kT/3$, but since we know that $E_0 \gg kT$ we must have $\tau \gg 2$, from which it follows that the temperature dependence of $\langle \sigma v \rangle$ must be determined essentially by the exponential factor. Thus the reaction rate increases rapidly with temperature, and this increase will be the greater the heavier the reacting nuclei. That is, the heavier the reacting nuclei, the greater the temperature sensitivity.

If any resonances lie close to E_0, i.e. fall within the interval $E_0 \pm \Delta/2$, then equation (E.31) will be quite inapplicable. We shall not consider here the changes that have to be made to take account of the presence of resonances (see, for example, Chapter 4 of the book by Clayton cited in the bibliography for Chapter VI), but note simply that they can lead to some reactions having significant rates at much lower temperatures than would otherwise be the case. At the same time it should be realised that the value of around 5 keV that we have deduced above for Δ in the case of hydrogen-burning stars is considerably smaller than the spacing of resonances in light nuclei (see, for example, figure 21.2), so that there will at least be *some* situations where equation (E.31) is valid.

APPENDIX F PARTIAL-WAVE ANALYSIS OF CROSS SECTION

Entrance channel wavefunction. In the formal development of this Appendix we describe nuclear reactions from the point of view of the entrance channel. That is, we distinguish between only two possibilities: either the incident particle returns to the entrance channel or the incident particle is *permanently* removed from the entrance channel. The former corresponds to elastic scattering, while all other processes are lumped together in the latter, and regarded simply as 'reaction', in the sense of the reaction cross section σ_{react} defined in equation (19.21). This is the point of view of the optical model (§20), although we stress that the present development is of much more general validity.

All that we have to consider, then, is the entrance channel wavefunction

$$\Psi = \chi_a \chi_A \psi(\boldsymbol{r}) \tag{F.1}$$

where χ_a and χ_A are the intrinsic wavefunctions of the incident and target nuclei and $\psi(\boldsymbol{r})$ the wavefunction of their relative motion. We are concerned only with the latter.

When the target and incident nuclei are separated beyond a certain distance R, the so-called *radius of interaction*, which is of the order of the sum of the radii of the two nuclei, $R_a + R_A$, all interaction between them, other than Coulomb, will cease. Then $\psi(\boldsymbol{r})$ will satisfy the Schrödinger equation

$$(\nabla^2 + k^2 - U_C)\psi(\boldsymbol{r}) = 0 \qquad (r > R) \tag{F.2}$$

where $U_C = (2\mu/\hbar^2)Z_a Z_A e^2/r$ represents the Coulomb interaction, non-vanishing in all cases except where the incident particle is a neutron (μ is the reduced mass of the system). For $r < R$, i.e. within the so-called *interaction region*, not only is (F.2) invalid but $\psi(\boldsymbol{r})$ is no longer even defined in general, since the interacting nuclei become deformed and lose their identity, so that the separation (F.1) breaks down.

Scattering amplitude and cross sections. To correspond to the actual set-up in a reaction experiment $\psi(r)$ must satisfy an asymptotic boundary condition of the form (20.34), i.e.

$$\psi(r) \sim e^{ikz} + \frac{f(\Omega)}{r} e^{ikr} \tag{F.3}$$

the first term representing the plane wave of the incident beam and the second the outgoing scattered wave. We shall now express the cross sections for elastic scattering and for reaction, this embracing all possibilities, in terms of the *scattering amplitude* $f(\Omega)$. (Here Ω represents a direction (θ, ϕ).)

We note first the quantum-mechanical expression for current density

$$j = -\frac{i\hbar}{2\mu}(\psi^* \nabla \psi - \psi \nabla \psi^*). \tag{F.4}$$

Then the total number of particles scattered into solid angle $d\Omega$ per unit time is

$$n(\Omega)\, d\Omega = -\frac{i\hbar}{2\mu}\left[\psi_{sc}^* \frac{\partial \psi_{sc}}{\partial r} - cc\right]_{r=D} D^2\, d\Omega \tag{F.5}$$

where ψ_{sc} is the second term of (F.3) and D is any value of r such that $kD \gg 1$, this guaranteeing the validity of (F.3). The plane-wave term is excluded because in practice the incident beam is shielded from the detector. Ignoring terms of the order of $(kD)^{-1}$, we find

$$n(\Omega)\, d\Omega = j_{inc} |f(\Omega)|^2\, d\Omega \tag{F.6}$$

where

$$j_{inc} = \frac{\hbar k}{\mu} = v_{inc} \tag{F.7}$$

is the current density of the incident beam (equal to the particle velocity v_{inc}, with the normalisation of (F.3)). Thus the differential elastic cross section is, from equation (19.19)

$$\frac{d\sigma_{el}}{d\Omega} = |f(\Omega)|^2 \tag{F.8}$$

so that the integrated elastic scattering cross section is

$$\sigma_{el} = \int_{4\pi} |f(\Omega)|^2\, d\Omega. \tag{F.9}$$

Since for us reaction is simply a question of permanent removal from the entrance channel, without regard to the products, the corresponding cross section is obtained simply by calculating the net inward flux of the entrance channel wavefunction $\psi(r)$ across a large sphere S centred on the target, and of radius D ($kD \gg 1$). From (F.4) we have for this flux

$$\mathcal{N} = \frac{i\hbar}{2\mu}\int_S \left(\psi^* \frac{\partial \psi}{\partial r} - \psi \frac{\partial \psi^*}{\partial r}\right) dS \tag{F.10}$$

where ψ is the *entire* wavefunction of asymptotic form (F.3). Writing this latter as

$$\psi \sim \exp(ikr\cos\theta) + \frac{f(\Omega)}{r} e^{ikr} \tag{F.11}$$

leads to

$$\mathcal{N} = -\frac{\hbar k}{\mu}\int_S \left(\cos\theta + \frac{|f(\Omega)|^2}{D^2} + \frac{1+\cos\theta}{D}\,\mathrm{Re}\, f(\Omega)\exp[ikD(1-\cos\theta)]\right) dS \tag{F.12}$$

where again we have dropped terms $O(1/kD)$. Using (F.7), (F.9) and the fact that the first term of the integrand vanishes on integrating over S, we have

$$\sigma_{\mathrm{react}} = -\sigma_{\mathrm{el}} - 2\pi D\,\mathrm{Re}\int_0^\pi (1+\cos\theta) f(\theta)\exp[ikD(1-\cos\theta)]\sin\theta\, d\theta \tag{F.13}$$

where we are now assuming axial symmetry. The integral here may be approximated as

$$\int_{-1}^{1} (1+\omega) f(\theta)\exp[ikD(1-\omega)]\, d\omega$$

$$\simeq f(\theta=0)\int_{-1}^{1} (1+\omega)\exp[ikD(1-\omega)]\, d\omega \tag{F.14}$$

since when kD is large the rapid oscillations of the exponential factor make the integrand effectively vanish everywhere except in the region of $\theta = 0$.But

$$\int_{-1}^{1} (1+\omega)\exp[ikD(1-\omega)]\, d\omega = \frac{2i}{kD} + \frac{1}{k^2D^2}[1-\exp(2ikD)]$$

$$\simeq \frac{2i}{kD} \tag{F.15}$$

the asymptotic limit corresponding to large kD. Then (F.13) becomes

$$\sigma_{\mathrm{react}} = -\sigma_{\mathrm{el}} + \frac{4\pi}{k}\,\mathrm{Im}\, f(\theta=0). \tag{F.16}$$

For the total process cross section, given as in equation (19.21), we now have the *optical theorem*

$$\sigma_{\mathrm{tot}} \equiv \sigma_{\mathrm{react}} + \sigma_{\mathrm{el}} = \frac{4\pi}{k}\,\mathrm{Im}\, f(\theta=0). \tag{F.17}$$

The remarkable fact that the total cross section is completely determined by the scattering amplitude in the forward direction alone is readily understood on realising that this is effectively the cross section for removal from the incident *beam* (not from the entrance channel), a process which may be regarded as arising from destructive interference between the scattered wave and the incident plane wave.

When a Coulomb force is acting the asymptotic form (F.3) should be modified by logarithmically-varying phase factors. However, these can be ignored in practice, since the Coulomb force between nuclei is always screened by atomic electrons. Thus no formal difficulty need arise and the relations (F.8), (F.9), (F.16) and (F.17) between cross sections and scattering amplitudes remain valid for reactions between charged particles.

Partial-wave analysis. The partial-wave expansion of the entrance channel wavefunction satisfying (F.2) beyond the interaction region takes, in the case of axial symmetry, the form

$$\psi = \frac{1}{kr}\sum_l u_l(r)P_l(\theta) \qquad (r > R) \tag{F.18}$$

where the radial function $u_l(r)$ satisfies

$$\left(\frac{\mathrm{d}^2}{\mathrm{d}r^2} + k^2 - \frac{l(l+1)}{r^2} - U_\mathrm{C}\right)u_l(r) = 0 \tag{F.19}$$

and the $P_l(\theta)$ are Legendre polynomials.

Any solution $u_l(r)$ of (F.19) may be expressed as some linear combination of any given pair of linearly-independent solutions of the same equation,

$$u_l(r) = A_l I_l(k, r) + B_l O_l(k, r) \tag{F.20}$$

where A_l and B_l are arbitrary constants, and we have introduced as solutions to (F.19) incoming and outgoing waves I_l and O_l, respectively. For the charge-free case these are

$$O_l(kr) = \mathrm{i}kr[j_l(kr) + \mathrm{i}n_l(kr)] \sim \exp \mathrm{i}\left(kr - \frac{l\pi}{2}\right) \tag{F.21a}$$

$$I_l(kr) = O_l^*(kr) \sim \exp\left[-\mathrm{i}\left(kr - \frac{l\pi}{2}\right)\right] \tag{F.21b}$$

where we have introduced the spherical Bessel and Neumann functions, $j_l(x)$ and $n_l(x)$, respectively†.

† See, for example, Merzbacher E 1970 *Quantum Mechanics* 2nd edn (New York: Wiley) pp. 195–6.

When the incident particles are charged (F.21) must be replaced by

$$O_l^{(C)}(k,r) = \mathrm{i}\,\mathrm{e}^{-\mathrm{i}\sigma_l}(F_l(n,kr) - \mathrm{i}G_l(n,kr)) \sim \exp \mathrm{i}\left(kr - \frac{l\pi}{2} - n\log 2kr\right) \tag{F.22a}$$

$$I_l^{(C)}(k,r) = O_l^{(C)*} \sim \exp\left[-\mathrm{i}\left(kr - \frac{l\pi}{2} - n\log 2kr\right)\right] \tag{F.22b}$$

where n is the Sommerfeld parameter of equation (18.6)

$$\sigma_l = \arg\Gamma(l+1+\mathrm{i}n) \tag{F.23}$$

and F_l and G_l are respectively the regular and irregular spherical Coulomb functions† having the following limiting behaviour for vanishing charge

$$\lim_{e\to 0} F_l(n,kr) = krj_l(kr) \qquad \lim_{e\to 0} G_l(n,kr) = krn_l(kr). \tag{F.24}$$

Actually, we have already remarked that atomic screening suppresses all asymptotic Coulomb effects, so that the logarithmically varying phase factors in (F.22) should be dropped. Thus

$$O_l^{(C)}(k,r) \sim O_l(kr) \qquad I_l^{(C)}(k,r) \sim I_l(kr). \tag{F.25}$$

In fact, the functions will be equal everywhere beyond the atomic screening. It is by invoking the physical fact of screening in this way that all the formal difficulties usually associated with the Coulomb force can be eliminated. On the other hand, the Coulomb force is still quantitatively important, since the screening has negligible effect in the vicinity of the nucleus, so that the Coulomb and charge-free solutions will be quite different here, and it is in this region that cross sections are determined, although they are measured in the asymptotic region, of course‡.

With the boundary condition (F.3) applicable to both incident neutrons and charged particles, the coefficients A_l in (F.20) are determined at once, since $\psi(r)$ can differ asymptotically (actually, beyond $r = R$ in the neutral case and beyond the atomic screening in the Coulomb case) from the plane wave with respect to outgoing waves only. Noting that

$$\mathrm{e}^{\mathrm{i}kz} = \frac{1}{2kr}\sum_l (2l+1)\mathrm{i}^{l+1}(I_l(kr) - O_l(kr))P_l(\theta) \tag{F.26}$$

† See, for example, Appendix B of the book of Preston and Bhaduri cited in the General Bibliography. Note, however, that the n_l of these authors differs by a sign with respect to ours.

‡ The only effect that screening has on cross sections is for scattering angles $<(k \times \text{atomic radius})^{-1}$, this being the angular radius of the first Fraunhofer diffraction minimum of the atom. But such angles will be so small in the nuclear context that measurements will not usually be possible, although it is precisely in this forward direction that the unscreened functions give rise to formal problems (singularities).

reduces (F.18) to

$$\psi = \frac{1}{2kr} \sum_l (2l+1) \mathrm{i}^{l+1} (I_l - U_l O_l) P_l(\theta) \qquad r > R \tag{F.27}$$

where the incoming and outgoing waves, I_l and O_l, respectively, are given by (F.21) or (F.22), as appropriate (in the Coulomb case we make explicit use of (F.25)). We have replaced B_l by $-\frac{1}{2}(2l+1)\mathrm{i}^{l+1}U_l$, where the unknown quantity U_l is called the *collision function* or *collision parameter*.

Rewriting (F.27) as

$$\psi \sim \mathrm{e}^{ikz} + \frac{1}{2kr} \sum_l (2l+1) \mathrm{i}^{l+1} (1 - U_l) O_l(kr) P_l(\theta) \tag{F.28}$$

(in the uncharged case this is valid everywhere beyond the radius of nuclear interaction, but in the charged case it can be valid only beyond the screening) and comparing with (F.3) gives a partial-wave expansion of the scattering amplitude

$$f(\theta) = \frac{\mathrm{i}}{2k} \sum_l (2l+1)(1 - U_l) P_l(\theta). \tag{F.29}$$

This now permits us to express cross sections directly in terms of the collision parameters. For elastic scattering (F.8) and (F.9) give us, respectively

$$\frac{\mathrm{d}\sigma_{\mathrm{el}}}{\mathrm{d}\Omega} = \frac{1}{4k^2} \left| \sum_l (2l+1)(1 - U_l) P_l(\theta) \right|^2 \tag{F.30}$$

and

$$\sigma_{\mathrm{el}} = \frac{\pi}{k^2} \sum_l (2l+1) \, | \, 1 - U_l \, |^2. \tag{F.31}$$

For the integrated, but not the differential, elastic scattering it is possible to define a partial-wave cross section

$$\sigma_{\mathrm{el}}^l = \frac{\pi}{k^2} (2l+1) \, | \, 1 - U_l \, |^2. \tag{F.32}$$

The total process cross section (F.17) reduces likewise to

$$\sigma_{\mathrm{tot}} = \frac{2\pi}{k^2} \sum_l (2l+1)(1 - \mathrm{Re}\, U_l) \tag{F.33}$$

which combined with (F.31) gives for the reaction cross section

$$\sigma_{\mathrm{react}} = \frac{\pi}{k^2} \sum_l (2l+1)(1 - | \, U_l \, |^2). \tag{F.34}$$

A partial-wave cross section may now be defined for each of these

$$\sigma^l_{\text{tot}} = \frac{2\pi}{k^2}(2l+1)(1 - \text{Re } U_l) \tag{F.35}$$

$$\sigma^l_{\text{react}} = \frac{\pi}{k^2}(2l+1)(1 - |U_l|^2). \tag{F.36}$$

Note, of course, that with our general definition of reaction processes, the concept of differential cross section is meaningless.

In the case of pure elastic scattering, i.e. no reaction possible, we have

$$u_l \sim \sin\left(kr - \frac{l\pi}{2} + \delta_l\right) = \frac{\mathrm{i}}{2}\,\mathrm{e}^{-\mathrm{i}\delta_l}(I_l - \mathrm{e}^{2\mathrm{i}\delta_l}O_l) \tag{F.37}$$

where δ_l is the phaseshift. Then in this case U_l takes the form

$$U_l = \mathrm{e}^{2\mathrm{i}\delta_l}. \tag{F.38}$$

The short range of purely nuclear interactions guarantees that only the first few partial waves contribute significantly to the various cross sections in the absence of Coulomb forces. On the other hand, equations (F.30) and (F.31) for elastic scattering are unpractical in the presence of Coulomb forces, since an infinite number of partial waves have to be taken into account. (Actually, screening will introduce an effective truncation, but the number of significant partial waves will still be inconveniently large.) We handle this difficulty by first considering the pure Coulomb case, i.e. no nuclear interaction and a point charge. With screening being taken into account the functions u_l will be modified regular Coulomb functions

$$\bar{F}_l = \frac{\mathrm{i}}{2}\,\mathrm{e}^{-\mathrm{i}\bar{\sigma}_l}(I_l^{(\mathrm{C})} - \mathrm{e}^{2\mathrm{i}\bar{\sigma}_l}O_l^{(\mathrm{C})}) \sim \sin\left(kr - \frac{l\pi}{2} + \bar{\sigma}_l\right). \tag{F.39}$$

Here, because of the screening, there is no longer a logarithmically varying phasefactor in the asymptotic form, and the $\bar{\sigma}_l$ will be slightly different in general from the σ_l of equation (F.23).

In any case, we see from (F.37) that the $\bar{\sigma}_l$ can be interpreted as phaseshifts in the Coulomb field, modified by screening. Then, with $U_l = \exp 2\mathrm{i}\bar{\sigma}_l$, equation (F.29) gives for the scattering amplitude

$$f^{\mathrm{C}}(\theta) = \frac{\mathrm{i}}{2k}\sum_l (2l+1)(1 - \mathrm{e}^{2\mathrm{i}\bar{\sigma}_l})P_l(\theta). \tag{F.40}$$

In the absence of screening, with $\bar{\sigma}_l$ given by equation (F.23), it is possible to express the sum of this infinite series in closed form, thus

$$f^{\mathrm{C}}(\theta) = -\frac{n}{2k}\operatorname{cosec}^2\frac{\theta}{2}\exp \mathrm{i}\left(2\sigma_0 - 2n\log\sin\frac{\theta}{2}\right) \tag{F.41}$$

an expression which leads directly to the Rutherford scattering formula

(2.17) for elastic scattering by a pure Coulomb force

$$\sigma_{\text{el}}^{\text{C}}(\theta) = \left(\frac{Z_1 Z_2 e^2}{4E}\right)^2 \operatorname{cosec}^4 \frac{\theta}{2}. \tag{F.42}$$

When screening is taken into account the singularity in the forward direction is removed, but it has negligible effect for all but the smallest scattering angles. Thus for all practical situations (F.41) may be used with impunity, and likewise (F.23) for $\bar{\sigma}_l$. The main role of screening is thus to remove the formal difficulties associated with Coulomb scattering.

When a nuclear interaction is superimposed† on the Coulomb force there will still be an infinite number of significant partial waves, but the series will in general no longer be summable. However, we can define a nuclear scattering amplitude f^{N} in terms of the total scattering amplitude f of (F.29) by

$$f(\theta) = f^{\text{N}}(\theta) + f^{\text{C}}(\theta). \tag{F.43}$$

Then from (F.40)

$$f^{\text{N}}(\theta) = \frac{\text{i}}{2k} \sum_l (2l+1)(\text{e}^{2\text{i}\sigma_l} - U_l) P_l(\theta). \tag{F.44}$$

Here only the first few partial waves contribute, beyond which $U_l = \exp 2\text{i}\sigma_l$, pure Coulomb scattering then being the only process possible. Thus, using the closed form (F.41) for $f^{\text{C}}(\theta)$ in (F.43), the problem of the excessive number of partial waves is solved.

Expressed in this way, the elastic scattering cross section contains three terms

$$\sigma_{\text{el}} = \sigma^{\text{C}} + \sigma_{\text{el}}^{\text{N}} + \sigma_{\text{el}}^{\text{NC}} \tag{F.45}$$

where σ^{C} is the pure Coulomb (Rutherford) cross section (F.42), $\sigma_{\text{el}}^{\text{N}}$ is a purely nuclear term given by (F.44), and $\sigma_{\text{el}}^{\text{NC}}$ represents interference between Coulomb and nuclear scattering. If we now write

$$U_l^{\text{N}} = \text{e}^{-2\text{i}\sigma_l} U_l \tag{F.46}$$

(F.44) becomes

$$f^{\text{N}}(\theta) = \frac{\text{i}}{2k} \sum_l (2l+1) \text{e}^{2\text{i}\sigma_l} (1 - U_l^{\text{N}}) P_l(\theta) \tag{F.47}$$

so that

$$\sigma_{\text{el}}^{\text{N}}(\theta) = \frac{1}{4k^2} \left| \sum_l (2l+1) \text{e}^{2\text{i}\sigma_l} (1 - U_l^{\text{N}}) P_l(\theta) \right|^2 \tag{F.48}$$

† This includes a modification of the Coulomb potential due to a finite nuclear charge distribution.

and

$$\sigma_{\text{el}}^{\text{N}} = \frac{\pi}{k^2} \sum_l (2l+1) \mid 1 - U_l^{\text{N}} \mid^2. \tag{F.49}$$

This latter is identical in form to (F.31).

For pure elastic scattering (F.38) and (F.46) give us

$$U_l^{\text{N}} = \text{e}^{2\text{i}\delta_l^{\text{N}}} \tag{F.50}$$

where we have introduced the nuclear phaseshift

$$\delta_l^{\text{N}} = \delta_l - \sigma_l \tag{F.51}$$

which represents the phaseshift induced by non-Coulomb forces, over and above the Coulomb phaseshift σ_l. Because it is defined effectively with respect to the function $F_l(n, kr)$, rather than $j_l(kr)$, this phaseshift will not be identical to the one given by the same nuclear force in the neutral case, unless the Coulomb barrier is negligible compared with the incident energy.

As for nuclear reactions, no problem arises, since only the first few partial waves contribute to the cross section (F.34). However, note that we can replace U_l in this latter by U_l^{N}.

Kinematical limits. The relations (F.30) to (F.36) parametrising cross sections in terms of the collision function are valid in all circumstances, no model having been assumed. We now obtain the following equally valid limits and inequalities.

In the first place, since a negative cross section is meaningless, it follows from (F.34) that

$$\mid U_l \mid \leqslant 1. \tag{F.52}$$

For reaction to take place we must have $\mid U_l \mid < 1$, in which case (F.31) shows that there will also be scattering. On the other hand, scattering without reaction is certainly possible.

Maximum reaction in a given partial wave occurs for $U_l = 0$

$$(\sigma_{\text{react}}^l)_{\text{max}} = \frac{\pi}{k^2}(2l+1) \tag{F.53}$$

which is simply equation (20.5).

APPENDIX G ATOMIC MASS TABLE

The following table is adapted from Wapstra A H and Audi G 1985 *Nucl. Phys.* **A432** 1–54, and is reproduced by kind permission of the North Holland Publishing Company and Professor A H Wapstra. The following abbreviations are used in the table:

EL element symbol
signifies estimated value.

See the text on pp. 20–21 for explanations of N, Z and A.

N	Z	A	EL	Mass excess (keV)
1	0	1	n	8071.369
0	1		H	7289.030
1	1	2	H	13135.824
2	1	3	H	14949.91
1	2		He	14931.32
3	1	4	H	25840
2	2		He	2424.92
1	3		Li	25120
3	2	5	He	11390
2	3		Li	11680
4	2	6	He	17592.3
3	3		Li	14085.6
2	4		Be	18374
5	2	7	He	26110
4	3		Li	14906.8
3	4		Be	15768.7
2	5		B	27870
6	2	8	He	31598
5	3		Li	20945.4
4	4		Be	4941.73
3	5		B	22920.3
2	6		C	35095
7	2	9	He	40810
6	3		Li	24953.9
5	4		Be	11347.7
4	5		B	12415.8
3	6		C	28913.2
7	3	10	Li	33830
6	4		Be	12607.0
5	5		B	12050.78
4	6		C	15701.7
3	7		N	39700#
8	3	11	Li	40900
7	4		Be	20174
6	5		B	8668.0
5	6		C	10650.1
4	7		N	24910
8	4	12	Be	25077
7	5		B	13369.5
6	6		C	.0
5	7		N	17338.1
4	8		O	32060
9	4	13	Be	34950#
8	5		B	16562.3
7	6		C	3125.025
6	7		N	5345.52
5	8		O	23111
10	4	14	Be	41020#
9	5		B	23664
8	6		C	3019.910
7	7		N	2863.436
6	8		O	8006.56
5	9		F	33610#
10	5	15	B	28970
9	6		C	9873.2
8	7		N	101.50
7	8		O	2855.5
6	9		F	16770
11	5	16	B	37640#
10	6		C	13694
9	7		N	5682.1
8	8		O	−4737.03
7	9		F	10680
6	10		Ne	23989
12	5	17	B	44010#
11	6		C	21030
10	7		N	7871
9	8		O	−809.3
8	9		F	1951.54
7	10		Ne	16480
12	6	18	C	24890
11	7		N	13117
10	8		O	−782.2
9	9		F	873.2
8	10		Ne	5319
7	11		Na	25320#
13	6	19	C	32760#
12	7		N	15873
11	8		O	3332.2
10	9		F	−1487.40
9	10		Ne	1751.0
8	11		Na	12929
14	6	20	C	38030#
13	7		N	22100#
12	8		O	3796.3
11	9		F	−17.33
10	10		Ne	−7046.2
9	11		Na	6841
8	12		Mg	17572
14	7	21	N	26050#
13	8		O	8130
12	9		F	−48
11	10		Ne	−5735.4
10	11		Na	−2188.6
9	12		Mg	10914
14	8	22	O	9440
13	9		F	2830
12	10		Ne	−8026.6
11	11		Na	−5184.6
10	12		Mg	−396.6
9	13		Al	18040
15	8	23	O	17460#
14	9		F	3350
13	10		Ne	−5155.5
12	11		Na	−9531.4
11	12		Mg	−5473.1
10	13		Al	6767
15	9	24	F	8750#
14	10		Ne	−5950
13	11		Na	−8419.5
12	12		Mg	−13933.1
11	13		Al	−55
10	14		Si	10755
16	9	25	F	12540#
15	10		Ne	−2160
14	11		Na	−9359
13	12		Mg	−13192.5
12	13		Al	−8915.4
11	14		Si	3827
16	10	26	Ne	440
15	11		Na	−6906
14	12		Mg	−16214.0
13	13		Al	−12209.9
12	14		Si	−7144
11	15		P	11260#
17	10	27	Ne	6750#
16	11		Na	−5650
15	12		Mg	−14586.2
14	13		Al	−17196.8
13	14		Si	−12385.3
12	15		P	−750
17	11	28	Na	−1140
16	12		Mg	−15018.8
15	13		Al	−16850.6
14	14		Si	−21492.4
13	15		P	−7161
12	16		S	4130
18	11	29	Na	2640
17	12		Mg	−10728
16	13		Al	−18215
15	14		Si	−21895.0
14	15		P	−16950.5
13	16		S	−3160
19	11	30	Na	8200
18	12		Mg	−9100
17	13		Al	−15890
16	14		Si	−24433.2
15	15		P	−20207.4
14	16		S	−14063
13	17		Cl	4840#
20	11	31	Na	11810
19	12		Mg	−3790#
18	13		Al	−15090
17	14		Si	−22950.2
16	15		P	−24440.7
15	16		S	−19045.2
14	17		Cl	−7070

N	Z	A	EL	Mass excess (keV)
21	11	32	Na	16530
20	12		Mg	−1770
19	13		Al	−11180#
18	14		Si	−24080.8
17	15		P	−24305.8
16	16		S	−26016.18
15	17		Cl	−13330
14	18		Ar	−2180
22	11	33	Na	21450
21	12		Mg	3930#
20	13		Al	−9270#
19	14		Si	−20570
18	15		P	−26338.0
17	16		S	−26586.51
16	17		Cl	−21003.9
15	18		Ar	−9380
23	11	34	Na	26640
22	12		Mg	6940#
21	13		Al	−4360#
20	14		Si	−19860
19	15		P	−24557.9
18	16		S	−29932.25
17	17		Cl	−24439.92
16	18		Ar	−18379
15	19		K	−1480#
23	12	35	Mg	13560#
22	13		Al	−1440#
21	14		Si	−14540#
20	15		P	−24940
19	16		S	−28846.89
18	17		Cl	−29013.72
17	18		Ar	−23048.8
16	19		K	−11168
23	13	36	Al	3910#
22	14		Si	−12760#
21	15		P	−20890
20	16		S	−30664.44
19	17		Cl	−29522.15
18	18		Ar	−30231.39
17	19		K	−17426
16	20		Ca	−6440
23	14	37	Si	−7000#
22	15		P	−19100#
21	16		S	−26896.59
20	17		Cl	−31761.75
19	18		Ar	−30947.9
18	19		K	−24799.4
17	20		Ca	−13160
24	14	38	Si	−4660#
23	15		P	−14660#
22	16		S	−26862
21	17		Cl	−29798.23
20	18		Ar	−34714.7
19	19		K	−28801.7
18	20		Ca	−22060
17	21		Sc	−4460#

N	Z	A	EL	Mass excess (keV)
24	15	39	P	−12300#
23	16		S	−23000#
22	17		Cl	−29804
21	18		Ar	−33242
20	19		K	−33806.6
19	20		Ca	−27276.0
18	21		Sc	−14180#
25	15	40	P	−7620#
24	16		S	−22520
23	17		Cl	−27540
22	18		Ar	−35039.6
21	19		K	−33534.8
20	20		Ca	−34846.9
19	21		Sc	−20527
18	22		Ti	−9064
25	16	41	S	−17870#
24	17		Cl	−27400
23	18		Ar	−33067.4
22	19		K	−35559.7
21	20		Ca	−35138.3
20	21		Sc	−28643.4
19	22		Ti	−15700
26	16	42	S	−16420#
25	17		Cl	−24420#
24	18		Ar	−34420
23	19		K	−35022.7
22	20		Ca	−38547.8
21	21		Sc	−32124.1
20	22		Ti	−25122
19	23		V	−8220#
26	17	43	Cl	−23130
25	18		Ar	−31980
24	19		K	−36592
23	20		Ca	−38409.4
22	21		Sc	−36188.7
21	22		Ti	−29321
20	23		V	−17920#
27	17	44	Cl	−20010#
26	18		Ar	−32262
25	19		K	−35810
24	20		Ca	−41469.9
23	21		Sc	−37815.4
22	22		Ti	−37549.1
21	23		V	−23800#
20	24		Cr	−13220
27	18	45	Ar	−29720
26	19		K	−36611
25	20		Ca	−40813.4
24	21		Sc	−41069.9
23	22		Ti	−39007.3
22	23		V	−31875
21	24		Cr	−19460

N	Z	A	EL	Mass excess (keV)
28	18	46	Ar	−29720
27	19		K	−35420
26	20		Ca	−43138
25	21		Sc	−41759.2
24	22		Ti	−44125.7
23	23		V	−37075.3
22	24		Cr	−29472
21	25		Mn	−12470#
28	19	47	K	−35698
27	20		Ca	−42343
26	21		Sc	−44331.3
25	22		Ti	−44931.9
24	23		V	−42004.8
23	24		Cr	−34554
22	25		Mn	−22650#
29	19	48	K	−32124
28	20		Ca	−44216
27	21		Sc	−44493
26	22		Ti	−48487.1
25	23		V	−44472
24	24		Cr	−42818
23	25		Mn	−29220#
30	19	49	K	−30790
29	20		Ca	−41291
28	21		Sc	−46555
27	22		Ti	−48558.1
26	23		V	−47956.3
25	24		Cr	−45328.9
24	25		Mn	−37611
23	26		Fe	−24470
30	20	50	Ca	−39571
29	21		Sc	−44538
28	22		Ti	−51426.2
27	23		V	−49219.7
26	24		Cr	−50257.9
25	25		Mn	−42626.0
24	26		Fe	−34470
31	20	51	Ca	−35940
30	21		Sc	−43220
29	22		Ti	−49727.2
28	23		V	−52199.7
27	24		Cr	−51448.3
26	25		Mn	−48239.7
25	26		Fe	−40218
24	27		Co	−27420#
31	21	52	Sc	−40040#
30	22		Ti	−49464
29	23		V	−51439.6
28	24		Cr	−55415.2
27	25		Mn	−50703.4
26	26		Fe	−48331
25	27		Co	−34300#

N	Z	A	EL	Mass excess (keV)
31	22	53	Ti	−46830
30	23		V	−51847
29	24		Cr	−55283.4
28	25		Mn	−54687.2
27	26		Fe	−50943.6
26	27		Co	−42640
25	28		Ni	−29410
32	22	54	Ti	−45430#
31	23		V	−49889
30	24		Cr	−56931.0
29	25		Mn	−55554.0
28	26		Fe	−56250.8
27	27		Co	−48009.3
26	28		Ni	−39210
32	23	55	V	−49150
31	24		Cr	−55105.9
30	25		Mn	−57709.2
29	26		Fe	−57477.4
28	27		Co	−54026.0
27	28		Ni	−45330
26	29		Cu	−31630#
33	23	56	V	−46110#
32	24		Cr	−55291
31	25		Mn	−56908.4
30	26		Fe	−60604.1
29	27		Co	−56038.0
28	28		Ni	−53902
27	29		Cu	−38500#
33	24	57	Cr	−52690#
32	25		Mn	−57488
31	26		Fe	−60178.9
30	27		Co	−59342.5
29	28		Ni	−56077.4
28	29		Cu	−47380#
27	30		Zn	−32610
34	24	58	Cr	−52050#
33	25		Mn	−55830
32	26		Fe	−62152.2
31	27		Co	−59844.3
30	28		Ni	−60225.1
29	29		Cu	−51662.4
28	30		Zn	−42210
34	25	59	Mn	−55477
33	26		Fe	−60661.9
32	27		Co	−62226.5
31	28		Ni	−61153.6
30	29		Cu	−56353.0
29	30		Zn	−47260
35	25	60	Mn	−52900
34	26		Fe	−61407
33	27		Co	−61647.1
32	28		Ni	−64470.7
31	29		Cu	−58343.8
30	30		Zn	−54185
35	26	61	Fe	−58919
34	27		Co	−62897.1
33	28		Ni	−64219.6
32	29		Cu	−61981.2
31	30		Zn	−56343
30	31		Ga	−47540#
36	26	62	Fe	−58896
35	27		Co	−61424
34	28		Ni	−66745.7
33	29		Cu	−62797
32	30		Zn	−61170
31	31		Ga	−51999
37	26	63	Fe	−55190
36	27		Co	−61839
35	28		Ni	−65512.8
34	29		Cu	−65578.7
33	30		Zn	−62211.6
32	31		Ga	−56690
31	32		Ge	−47390#
37	27	64	Co	−59791
36	28		Ni	−67098.0
35	29		Cu	−65423.5
34	30		Zn	−66001.7
33	31		Ga	−58837
32	32		Ge	−54430
38	27	65	Co	−59160
37	28		Ni	−65124.8
36	29		Cu	−67261.0
35	30		Zn	−65910.2
34	31		Ga	−62654.4
33	32		Ge	−56410
32	33		As	−47310#
38	28	66	Ni	−66028
37	29		Cu	−66255.6
36	30		Zn	−68898.7
35	31		Ga	−63724
34	32		Ge	−61622
33	33		As	−52070
39	28	67	Ni	−63742
38	29		Cu	−67303
37	30		Zn	−67879.3
36	31		Ga	−66878.4
35	32		Ge	−62656
34	33		As	−56650
33	34		Se	−46860#
40	28	68	Ni	−63482
39	29		Cu	−65560
38	30		Zn	−70006.1
37	31		Ga	−67085.0
36	32		Ge	−66978
35	33		As	−58880
34	34		Se	−54080#
41	28	69	Ni	−60460
40	29		Cu	−65741
39	30		Zn	−68417.0
38	31		Ga	−69322.5
37	32		Ge	−67097
36	33		As	−63080
35	34		Se	−56290
34	35		Br	−46790#
41	29	70	Cu	−62982
40	30		Zn	−69560
39	31		Ga	−68905
38	32		Ge	−70561.5
37	33		As	−64340
36	34		Se	−61590
35	35		Br	−51190#
42	29	71	Cu	−62820#
41	30		Zn	−67322
40	31		Ga	−70141.5
39	32		Ge	−69905.9
38	33		As	−67893
37	34		Se	−63090#
36	35		Br	−56590#
35	36		Kr	−46490#
42	30	72	Zn	−68134
41	31		Ga	−68591.2
40	32		Ge	−72583.6
39	33		As	−68228
38	34		Se	−67897
37	35		Br	−59030#
36	36		Kr	−53970#
43	30	73	Zn	−65410
42	31		Ga	−69705
41	32		Ge	−71294.7
40	33		As	−70955
39	34		Se	−68215
38	35		Br	−63640
37	36		Kr	−56890
36	37		Rb	−46590#
44	30	74	Zn	−65707
43	31		Ga	−68060
42	32		Ge	−73423.4
41	33		As	−70861.1
40	34		Se	−72215.0
39	35		Br	−65300
38	36		Kr	−62140
37	37		Rb	−51750
45	30	75	Zn	−62700
44	31		Ga	−68466
43	32		Ge	−71857.5
42	33		As	−73035.1
41	34		Se	−72171.3
40	35		Br	−69161
39	36		Kr	−64246
38	37		Rb	−57280

N	Z	A	EL	Mass excess (keV)
46	30	76	Zn	-62460
45	31		Ga	-66440
44	32		Ge	-73214.5
43	33		As	-72290.7
42	34		Se	-75254.1
41	35		Br	-70302
40	36		Kr	-68969
39	37		Rb	-60580
47	30	77	Zn	-58910#
46	31		Ga	-66410#
45	32		Ge	-71215.4
44	33		As	-73918.5
43	34		Se	-74601.6
42	35		Br	-73237
41	36		Kr	-70227
40	37		Rb	-64950
39	38		Sr	-57890
48	30	78	Zn	-57960#
47	31		Ga	-63560#
46	32		Ge	-71863
45	33		As	-72816
44	34		Se	-77028.1
43	35		Br	-73454
42	36		Kr	-74151
41	37		Rb	-66980
40	38		Sr	-63650#
48	31	79	Ga	-62760
47	32		Ge	-69530
46	33		As	-73639
45	34		Se	-75919.1
44	35		Br	-76070.0
43	36		Kr	-74442
42	37		Rb	-70837
41	38		Sr	-65340#
40	39		Y	-58240#
49	31	80	Ga	-59380
48	32		Ge	-69380
47	33		As	-72165
46	34		Se	-77762.1
45	35		Br	-75891.0
44	36		Kr	-77892
43	37		Rb	-72173
42	38		Sr	-70190
41	39		Y	-61190#
50	31	81	Ga	-57990
49	32		Ge	-66310
48	33		As	-72535
47	34		Se	-76391.8
46	35		Br	-77977
45	36		Kr	-77697
44	37		Rb	-75461
43	38		Sr	-71470
42	39		Y	-65950
41	40		Zr	-58790

N	Z	A	EL	Mass excess (keV)
50	32	82	Ge	-65380
49	33		As	-70078
48	34		Se	-77596.1
47	35		Br	-77499
46	36		Kr	-80591
45	37		Rb	-76202
44	38		Sr	-75997
43	39		Y	-68180
42	40		Zr	-64180
51	32	83	Ge	-61240#
50	33		As	-69880
49	34		Se	-75343
48	35		Br	-79011
47	36		Kr	-79983
46	37		Rb	-79044
45	38		Sr	-76788
44	39		Y	-72380
43	40		Zr	-66360
51	33	84	As	-66080#
50	34		Se	-75952
49	35		Br	-77778
48	36		Kr	-82431
47	37		Rb	-79746
46	38		Sr	-80640
45	39		Y	-74230
44	40		Zr	-71430#
52	33	85	As	-63510#
51	34		Se	-72420
50	35		Br	-78607
49	36		Kr	-81477
48	37		Rb	-82164.4
47	38		Sr	-81099
46	39		Y	-77839
45	40		Zr	-73150
44	41		Nb	-66740#
52	34	86	Se	-70540
51	35		Br	-75640
50	36		Kr	-83262
49	37		Rb	-82743.4
48	38		Sr	-84517.7
47	39		Y	-79278
46	40		Zr	-77980#
45	41		Nb	-69580#
53	34	87	Se	-66710#
52	35		Br	-73880
51	36		Kr	-80706
50	37		Rb	-84592.3
49	38		Sr	-84874.6
48	39		Y	-83013.6
47	40		Zr	-79348
46	41		Nb	-74180
45	42		Mo	-67440

N	Z	A	EL	Mass excess (keV)
53	35	88	Br	-70720
52	36		Kr	-79687
51	37		Rb	-82600
50	38		Sr	-87916.2
49	39		Y	-84294
48	40		Zr	-83626
47	41		Nb	-76430#
46	42		Mo	-72830#
54	35	89	Br	-68420#
53	36		Kr	-76720
52	37		Rb	-81713
51	38		Sr	-86210
50	39		Y	-87702.2
49	40		Zr	-84869
48	41		Nb	-80622
47	42		Mo	-75004
46	43		Tc	-68000#
55	35	90	Br	-64260#
54	36		Kr	-74960
53	37		Rb	-79353
52	38		Sr	-85942
51	39		Y	-86488.1
50	40		Zr	-88769.7
49	41		Nb	-82659
48	42		Mo	-80172
47	43		Tc	-70970#
55	36	91	Kr	-71370
54	37		Rb	-77794
53	38		Sr	-83661
52	39		Y	-86347
51	40		Zr	-87892.8
50	41		Nb	-86638
49	42		Mo	-82200
48	43		Tc	-75980
47	44		Ru	-68400#
56	36	92	Kr	-68680
55	37		Rb	-74836
54	38		Sr	-82956
53	39		Y	-84844
52	40		Zr	-88456.6
51	41		Nb	-86450.7
50	42		Mo	-86808
49	43		Tc	-78938
48	44		Ru	-74410#
57	36	93	Kr	-64150
56	37		Rb	-72679
55	38		Sr	-80121
54	39		Y	-84235
53	40		Zr	-87119.3
52	41		Nb	-87209.8
51	42		Mo	-86804
50	43		Tc	-83606
49	44		Ru	-77270
48	45		Rh	-69110#

N	Z	A	EL	Mass excess (keV)
57	37	94	Rb	-68529
56	38		Sr	-78836
55	39		Y	-82348
54	40		Zr	-87267.9
53	41		Nb	-86367.9
52	42		Mo	-88413.2
51	43		Tc	-84157
50	44		Ru	-82567
49	45		Rh	-72970#
58	37	95	Rb	-65808
57	38		Sr	-75090
56	39		Y	-81214
55	40		Zr	-85659.3
54	41		Nb	-86783.6
53	42		Mo	-87709.2
52	43		Tc	-86018
51	44		Ru	-83449
50	45		Rh	-78340
49	46		Pd	-70150#
59	37	96	Rb	-61140
58	38		Sr	-72890
57	39		Y	-78300
56	40		Zr	-85442
55	41		Nb	-85605
54	42		Mo	-88792.1
53	43		Tc	-85819
52	44		Ru	-86071
51	45		Rh	-79630
50	46		Pd	-76370#
60	37	97	Rb	-58280
59	38		Sr	-68800
58	39		Y	-76270
57	40		Zr	-82950
56	41		Nb	-85608.3
55	42		Mo	-87542.1
54	43		Tc	-87222
53	44		Ru	-86111
52	45		Rh	-82600
51	46		Pd	-77800
50	47		Ag	-70900#
61	37	98	Rb	-54060
60	38		Sr	-66490
59	39		Y	-72370
58	40		Zr	-81288
57	41		Nb	-83528
56	42		Mo	-88113.2
55	43		Tc	-86429
54	44		Ru	-88225
53	45		Rh	-83168
52	46		Pd	-81299
51	47		Ag	-73070#
62	37	99	Rb	-50860
61	38		Sr	-62180
60	39		Y	-70130
59	40		Zr	-77740
58	41		Nb	-82327
57	42		Mo	-85967.4
56	43		Tc	-87324.4
55	44		Ru	-87618.0
54	45		Rh	-85519
53	46		Pd	-82192
52	47		Ag	-76760
51	48		Cd	-69990#
62	38	100	Sr	-60020#
61	39		Y	-66720#
60	40		Zr	-76620
59	41		Nb	-79950
58	42		Mo	-86186
57	43		Tc	-86017.4
56	44		Ru	-89219.9
55	45		Rh	-85590
54	46		Pd	-85207
53	47		Ag	-78120
52	48		Cd	-74310#
62	39	101	Y	-64380#
61	40		Zr	-73100
60	41		Nb	-78880
59	42		Mo	-83513
58	43		Tc	-86325
57	44		Ru	-87950.6
56	45		Rh	-87413
55	46		Pd	-85431
54	47		Ag	-81220
53	48		Cd	-75690
52	49		In	-68410#
62	40	102	Zr	-71760#
61	41		Nb	-76350
60	42		Mo	-83559
59	43		Tc	-84573
58	44		Ru	-89099.5
57	45		Rh	-86803
56	46		Pd	-87902
55	47		Ag	-82020
54	48		Cd	-79700#
53	49		In	-70420#
52	50		Sn	-64800#
63	40	103	Zr	-67610#
62	41		Nb	-75110#
61	42		Mo	-80610#
60	43		Tc	-84606
59	44		Ru	-87260.6
58	45		Rh	-88027
57	46		Pd	-87455
56	47		Ag	-84780
55	48		Cd	-80620
54	49		In	-74420#
53	50		Sn	-66920#
63	41	104	Nb	-71780#
62	42		Mo	-80480#
61	43		Tc	-82480
60	44		Ru	-88098
59	45		Rh	-86954
58	46		Pd	-89397
57	47		Ag	-85118
56	48		Cd	-83974
55	49		In	-75970#
54	50		Sn	-71470#
53	51		Sb	-59270#
64	41	105	Nb	-70140#
63	42		Mo	-77140#
62	43		Tc	-82140#
61	44		Ru	-85937
60	45		Rh	-87853
59	46		Pd	-88419
58	47		Ag	-87076
57	48		Cd	-84339
56	49		In	-79589
55	50		Sn	-73270
54	51		Sb	-64090#
64	42	106	Mo	-76430#
63	43		Tc	-79630#
62	44		Ru	-86330
61	45		Rh	-86370
60	46		Pd	-89910
59	47		Ag	-86944
58	48		Cd	-87132
57	49		In	-80590
56	50		Sn	-77290#
55	51		Sb	-66490#
54	52		Te	-58050#
65	42	107	Mo	-72510#
64	43		Tc	-78960#
63	44		Ru	-83710
62	45		Rh	-86862
61	46		Pd	-88374
60	47		Ag	-88407
59	48		Cd	-86990
58	49		In	-83570
57	50		Sn	-78370#
56	51		Sb	-70670#
55	52		Te	-60510#
65	43	108	Tc	-75990#
64	44		Ru	-83700
63	45		Rh	-85090
62	46		Pd	-89522
61	47		Ag	-87605
60	48		Cd	-89260
59	49		In	-84135
58	50		Sn	-82090
57	51		Sb	-72530#
56	52		Te	-65620#
55	53		I	-52550#

N	Z	A	EL	Mass excess (keV)
66	43	109	Tc	−74910#
65	44		Ru	−80810#
64	45		Rh	−85014
63	46		Pd	−87604
62	47		Ag	−88720
61	48		Cd	−88536
60	49		In	−86505
59	50		Sn	−82630
58	51		Sb	−76250
57	52		Te	−67650
56	53		I	−57760#
66	44	110	Ru	−80340#
65	45		Rh	−82940
64	46		Pd	−88337
63	47		Ag	−87458
62	48		Cd	−90351
61	49		In	−86410
60	50		Sn	−85830
59	51		Sb	−77530#
58	52		Te	−72140#
57	53		I	−60490#
56	54		Xe	−51750#
67	44	111	Ru	−76920#
66	45		Rh	−82320#
65	46		Pd	−86020
64	47		Ag	−88218
63	48		Cd	−89255
62	49		In	−88391
61	50		Sn	−85939
60	51		Sb	−80840#
59	52		Te	−73470
58	53		I	−64970#
57	54		Xe	−54380#
67	45	112	Rh	−79730#
66	46		Pd	−86329
65	47		Ag	−86623
64	48		Cd	−90582.3
63	49		In	−87994
62	50		Sn	−88654
61	51		Sb	−81589
60	52		Te	−77300
59	53		I	−67120#
58	54		Xe	−59880#
68	45	113	Rh	−78840#
67	46		Pd	−83680
66	47		Ag	−87041
65	48		Cd	−89051.1
64	49		In	−89367
63	50		Sn	−88328
62	51		Sb	−84421
61	52		Te	−78320#
60	53		I	−71120
59	54		Xe	−62130
58	55		Cs	−51610#
68	46	114	Pd	−83540
67	47		Ag	−84990
66	48		Cd	−90022.9
65	49		In	−88570
64	50		Sn	−90557
63	51		Sb	−84670
62	52		Te	−81760#
61	53		I	−72860#
60	54		Xe	−66910#
59	55		Cs	−54710#
69	46	115	Pd	−80490#
68	47		Ag	−84950
67	48		Cd	−88092.4
66	49		In	−89534
65	50		Sn	−90032
64	51		Sb	−87002
63	52		Te	−82250
62	53		I	−76300#
61	54		Xe	−68670#
60	55		Cs	−59550#
70	46	116	Pd	−80110
69	47		Ag	−82720
68	48		Cd	−88721
67	49		In	−88247
66	50		Sn	−91523
65	51		Sb	−86816
64	52		Te	−85280
63	53		I	−77520
62	54		Xe	−73020#
61	55		Cs	−62300
70	47	117	Ag	−82250
69	48		Cd	−86417
68	49		In	−88943
67	50		Sn	−90396.4
66	51		Sb	−88641
65	52		Te	−85110
64	53		I	−80610#
63	54		Xe	−74290#
62	55		Cs	−66230
61	56		Ba	−56930#
71	47	118	Ag	−79580
70	48		Cd	−86710
69	49		In	−87450
68	50		Sn	−91651.6
67	51		Sb	−87995
66	52		Te	−87647
65	53		I	−81250#
64	54		Xe	−78050#
63	55		Cs	−68240
62	56		Ba	−61950#
72	47	119	Ag	−78590
71	48		Cd	−83940
70	49		In	−87730
69	50		Sn	−90065.7
68	51		Sb	−89472
67	52		Te	−87178
66	53		I	−83810
65	54		Xe	−78820
64	55		Cs	−72200
63	56		Ba	−63960#
73	47	120	Ag	−75770
72	48		Cd	−83973
71	49		In	−85800
70	50		Sn	−91101.7
69	51		Sb	−88421
68	52		Te	−89380
67	53		I	−83980
66	54		Xe	−82030
65	55		Cs	−73770
64	56		Ba	−68470#
74	47	121	Ag	−74550
73	48		Cd	−80950
72	49		In	−85840
71	50		Sn	−89201.8
70	51		Sb	−89590.7
69	52		Te	−88542
68	53		I	−86263
67	54		Xe	−82490
66	55		Cs	−77090
65	56		Ba	−70140#
74	48	122	Cd	−80580#
73	49		In	−83580
72	50		Sn	−89945.4
71	51		Sb	−88325.9
70	52		Te	−90309
69	53		I	−86075
68	54		Xe	−85540
67	55		Cs	−78160
66	56		Ba	−74360#
75	48	123	Cd	−77320#
74	49		In	−83420
73	50		Sn	−87820.2
72	51		Sb	−89222.9
71	52		Te	−89171.7
70	53		I	−87939
69	54		Xe	−85261
68	55		Cs	−81050
67	56		Ba	−75260#
75	49	124	In	−81060
74	50		Sn	−88237.1
73	51		Sb	−87619.0
72	52		Te	−90525.2
71	53		I	−87368
70	54		Xe	−87659.6
69	55		Cs	−81720
68	56		Ba	−78820#
76	49	125	In	−80420
75	50		Sn	−85898.2
74	51		Sb	−88258
73	52		Te	−89025.0
72	53		I	−88846.8
71	54		Xe	−87191.5
70	55		Cs	−84092
69	56		Ba	−79510

N	Z	A	EL	Mass excess (keV)
77	49	126	In	−77810
76	50		Sn	−86021
75	51		Sb	−86400
74	52		Te	−90067.3
73	53		I	−87912
72	54		Xe	−89162
71	55		Cs	−84334
70	56		Ba	−82660#
78	49	127	In	−77010
77	50		Sn	−83504
76	51		Sb	−86705
75	52		Te	−88286
74	53		I	−88984
73	54		Xe	−88323
72	55		Cs	−86231
71	56		Ba	−82780
70	57		La	−77980#
79	49	128	In	−74000
78	50		Sn	−83310
77	51		Sb	−84600
76	52		Te	−88993
75	53		I	−87738
74	54		Xe	−89860.8
73	55		Cs	−85926
72	56		Ba	−85478
71	57		La	−78880
80	49	129	In	−73030
79	50		Sn	−80630
78	51		Sb	−84631
77	52		Te	−87008
76	53		I	−88506
75	54		Xe	−88697.4
74	55		Cs	−87536
73	56		Ba	−85100
72	57		La	−81380
81	49	130	In	−69990#
80	50		Sn	−80190
79	51		Sb	−82360
78	52		Te	−87348
77	53		I	−86897
76	54		Xe	−89881.1
75	55		Cs	−86859
74	56		Ba	−87299
73	57		La	−81600#
72	58		Ce	−79400#
82	49	131	In	−68550
81	50		Sn	−77370
80	51		Sb	−82020
79	52		Te	−85206
78	53		I	−87455
77	54		Xe	−88426
76	55		Cs	−88079
75	56		Ba	−86721
74	57		La	−83760
73	58		Ce	−79860#
82	50	132	Sn	−76610
81	51		Sb	−79730
80	52		Te	−85217
79	53		I	−85710
78	54		Xe	−89290
77	55		Cs	−87160
76	56		Ba	−88453
75	57		La	−83740
74	58		Ce	−82440#
73	59		Pr	−75340#
83	50	133	Sn	−69990#
82	51		Sb	−79040
81	52		Te	−82990
80	53		I	−85910
79	54		Xe	−87665
78	55		Cs	−88093
77	56		Ba	−87572
76	57		La	−85570#
75	58		Ce	−82570#
74	59		Pr	−78070#
83	51	134	Sb	−74000
82	52		Te	−82410
81	53		I	−83970
80	54		Xe	−88125
79	55		Cs	−86913
78	56		Ba	−88972
77	57		La	−85270
76	58		Ce	−84870#
75	59		Pr	−78770#
74	60		Nd	−75570#
84	51	135	Sb	−70310#
83	52		Te	−77850
82	53		I	−83813
81	54		Xe	−86509
80	55		Cs	−87668
79	56		Ba	−87873
78	57		La	−86673
77	58		Ce	−84657
76	59		Pr	−80910
75	60		Nd	−76210#
84	52	136	Te	−74410
83	53		I	−79510
82	54		Xe	−86431
81	55		Cs	−86361
80	56		Ba	−88909
79	57		La	−86040
78	58		Ce	−86500
77	59		Pr	−81380
76	60		Nd	−79170
75	61		Pm	−71280#
85	52	137	Te	−69480#
84	53		I	−76500
83	54		Xe	−82385
82	55		Cs	−86561
81	56		Ba	−87736
80	57		La	−87130
79	58		Ce	−85910
78	59		Pr	−83200
77	60		Nd	−79400
76	61		Pm	−74100#
85	53	138	I	−72310
84	54		Xe	−80130
83	55		Cs	−82900
82	56		Ba	−88276
81	57		La	−86531
80	58		Ce	−87575
79	59		Pr	−83138
78	60		Nd	−82140#
77	61		Pm	−75140#
76	62		Sm	−71340#
86	53	139	I	−68880
85	54		Xe	−75700
84	55		Cs	−80715
83	56		Ba	−84928
82	57		La	−87238
81	58		Ce	−86973
80	59		Pr	−84844
79	60		Nd	−82040
78	61		Pm	−77520
77	62		Sm	−72090
86	54	140	Xe	−73020
85	55		Cs	−77076
84	56		Ba	−83294
83	57		La	−84328
82	58		Ce	−88089
81	59		Pr	−84701
80	60		Nd	−84481
79	61		Pm	−78410
78	62		Sm	−75410#
77	63		Eu	−67210#
87	54	141	Xe	−68360
86	55		Cs	−74515
85	56		Ba	−79771
84	57		La	−83000
83	58		Ce	−85446
82	59		Pr	−86027
81	60		Nd	−84213
80	61		Pm	−80480
79	62		Sm	−75942
78	63		Eu	−69980
88	54	142	Xe	−65550
87	55		Cs	−70590
86	56		Ba	−77910
85	57		La	−80025
84	58		Ce	−84542
83	59		Pr	−83799
82	60		Nd	−85959
81	61		Pm	−81070
80	62		Sm	−78986
79	63		Eu	−71590
78	64		Gd	−67190#
88	55	143	Cs	−67790
87	56		Ba	−74070
86	57		La	−78320
85	58		Ce	−81615
84	59		Pr	−83077
83	60		Nd	−84012
82	61		Pm	−82969
81	62		Sm	−79526
80	63		Eu	−74380
79	64		Gd	−68480#

N	Z	A	EL	Mass excess (keV)
89	55	144	Cs	−63410
88	56		Ba	−71870
87	57		La	−74850
86	58		Ce	−80442
85	59		Pr	−80760
84	60		Nd	−83757
83	61		Pm	−81424
82	62		Sm	−81974
81	63		Eu	−75645
80	64		Gd	−71940#
79	65		Tb	−62940#
90	55	145	Cs	−60240
89	56		Ba	−68040
88	57		La	−72990
87	58		Ce	−77100
86	59		Pr	−79636
85	60		Nd	−81441
84	61		Pm	−81280
83	62		Sm	−80660
82	63		Eu	−77998
81	64		Gd	−72950
80	65		Tb	−66200#
91	55	146	Cs	−55690
90	56		Ba	−65100
89	57		La	−69370
88	58		Ce	−75760
87	59		Pr	−76780
86	60		Nd	−80935
85	61		Pm	−79450
84	62		Sm	−80992
83	63		Eu	−77114
82	64		Gd	−76100
81	65		Tb	−67860
80	66		Dy	−62860#
92	55	147	Cs	−52380
91	56		Ba	−61260#
90	57		La	−66970#
89	58		Ce	−72160
88	59		Pr	−75470
87	60		Nd	−78156
86	61		Pm	−79052
85	62		Sm	−79276
84	63		Eu	−77555
83	64		Gd	−75505
82	65		Tb	−70960
81	66		Dy	−64570
93	55	148	Cs	−47590
92	56		Ba	−58510#
91	57		La	−63910#
90	58		Ce	−70410
89	59		Pr	−72460
88	60		Nd	−77418
87	61		Pm	−76874
86	62		Sm	−79346
85	63		Eu	−76266
84	64		Gd	−76278
83	65		Tb	−70670
82	66		Dy	−67980
81	67		Ho	−58370#
93	56	149	Ba	−53390#
92	57		La	−61190#
91	58		Ce	−67290#
90	59		Pr	−70988
89	60		Nd	−74385
88	61		Pm	−76074
87	62		Sm	−77147
86	63		Eu	−76452
85	64		Gd	−75131
84	65		Tb	−71495
83	66		Dy	−67890#
82	67		Ho	−61850#
93	57	150	La	−57890#
92	58		Ce	−65510#
91	59		Pr	−68590#
90	60		Nd	−73694
89	61		Pm	−73607
88	62		Sm	−77061
87	63		Eu	−74798
86	64		Gd	−75766
85	65		Tb	−71102
84	66		Dy	−69325
83	67		Ho	−62220#
82	68		Er	−58020#
93	58	151	Ce	−62260#
92	59		Pr	−67160#
91	60		Nd	−70957
90	61		Pm	−73400
89	62		Sm	−74587
88	63		Eu	−74663
87	64		Gd	−74198
86	65		Tb	−71632
85	66		Dy	−68902
84	67		Ho	−63803
83	68		Er	−58500#
82	69		Tm	−51000#
93	59	152	Pr	−64560#
92	60		Nd	−70160
91	61		Pm	−71270
90	62		Sm	−74773
89	63		Eu	−72897
88	64		Gd	−74719
87	65		Tb	−70869
86	66		Dy	−70127
85	67		Ho	−63740
84	68		Er	−60620
83	69		Tm	−51740#
82	70		Yb	−46320#
93	60	153	Nd	−67370#
92	61		Pm	−70669
91	62		Sm	−72569
90	63		Eu	−73379
89	64		Gd	−72895
88	65		Tb	−71316
87	66		Dy	−69146
86	67		Ho	−65023
85	68		Er	−60670#
84	69		Tm	−54180#
83	70		Yb	−47270#
82	71		Lu	−38380#
94	60	154	Nd	−65770#
93	61		Pm	−68470
92	62		Sm	−72466
91	63		Eu	−71749
90	64		Gd	−73718
89	65		Tb	−70160
88	66		Dy	−70395
87	67		Ho	−64637
86	68		Er	−62623
85	69		Tm	−54630#
84	70		Yb	−50120#
83	71		Lu	−39630#
82	72		Hf	−32820#
94	61	155	Pm	−67100#
93	62		Sm	−70202
92	63		Eu	−71829
91	64		Gd	−72082
90	65		Tb	−71260
89	66		Dy	−69166
88	67		Ho	−66064
87	68		Er	−62360
86	69		Tm	−56810
85	70		Yb	−50740#
84	71		Lu	−42770#
83	72		Hf	−34440#
95	61	156	Pm	−64480#
94	62		Sm	−69380
93	63		Eu	−70094
92	64		Gd	−72547
91	65		Tb	−70103
90	66		Dy	−70536
89	67		Ho	−65540#
88	68		Er	−64000#
87	69		Tm	−56970
86	70		Yb	−53380
85	71		Lu	−43720#
84	72		Hf	−37860#
83	73		Ta	−25920#
95	62	157	Sm	−66870
94	63		Eu	−69473
93	64		Gd	−70835
92	65		Tb	−70773
91	66		Dy	−69434
90	67		Ho	−66890
89	68		Er	−63420
88	69		Tm	−58790#
87	70		Yb	−53620#
86	71		Lu	−46630#
85	72		Hf	−38960#
84	73		Ta	−29580#

N	Z	A	EL	Mass excess (keV)
96	62	158	Sm	−65200#
95	63		Eu	−67250
94	64		Gd	−70702
93	65		Tb	−69480
92	66		Dy	−70419
91	67		Ho	−66200
90	68		Er	−65200#
89	69		Tm	−58700#
88	70		Yb	−56023
87	71		Lu	−47240
86	72		Hf	−42300#
85	73		Ta	−31000#
84	74		W	−23780#
96	63	159	Eu	−66059
95	64		Gd	−68573
94	65		Tb	−69544
93	66		Dy	−69178
92	67		Ho	−67342
91	68		Er	−64573
90	69		Tm	−60570#
89	70		Yb	−55930#
88	71		Lu	−49850
87	72		Hf	−43090#
86	73		Ta	−34600#
85	74		W	−25550#
97	63	160	Eu	−63450#
96	64		Gd	−67954
95	65		Tb	−67848
94	66		Dy	−69683
93	67		Ho	−66397
92	68		Er	−66063
91	69		Tm	−60460#
90	70		Yb	−58060#
89	71		Lu	−50260#
88	72		Hf	−46060
87	73		Ta	−35740#
86	74		W	−29360#
97	64	161	Gd	−65518
96	65		Tb	−67473
95	66		Dy	−68065
94	67		Ho	−67208
93	68		Er	−65209
92	69		Tm	−62010#
91	70		Yb	−57810#
90	71		Lu	−52510#
89	72		Hf	−46480#
88	73		Ta	−38920#
87	74		W	−30610#
86	75		Re	−20710#
98	64	162	Gd	−64260
97	65		Tb	−65660
96	66		Dy	−68190
95	67		Ho	−66051
94	68		Er	−66347
93	69		Tm	−61560
92	70		Yb	−59750#
91	71		Lu	−52660#
90	72		Hf	−49179
89	73		Ta	−39910#
88	74		W	−34200#
87	75		Re	−22300#
98	65	163	Tb	−64690
97	66		Dy	−66390
96	67		Ho	−66387
95	68		Er	−65177
94	69		Tm	−62738
93	70		Yb	−59370
92	71		Lu	−54740#
91	72		Hf	−49390#
90	73		Ta	−42690#
89	74		W	−35150#
88	75		Re	−26110#
87	76		Os	−16450#
99	65	164	Tb	−62120
98	66		Dy	−65977
97	67		Ho	−64939
96	68		Er	−65952
95	69		Tm	−61990
94	70		Yb	−60990#
93	71		Lu	−54690#
92	72		Hf	−51790#
91	73		Ta	−43440#
90	74		W	−38360
89	75		Re	−27390#
88	76		Os	−20460#
99	66	165	Dy	−63622
98	67		Ho	−64908
97	68		Er	−64531
96	69		Tm	−62939
95	70		Yb	−60176
94	71		Lu	−56380#
93	72		Hf	−51650#
92	73		Ta	−45880#
91	74		W	−39020#
90	75		Re	−30840#
89	76		Os	−21870#
100	66	166	Dy	−62594
99	67		Ho	−63081
98	68		Er	−64935
97	69		Tm	−61888
96	70		Yb	−61595
95	71		Lu	−56120
94	72		Hf	−53790#
93	73		Ta	−46310#
92	74		W	−41899
91	75		Re	−31910#
90	76		Os	−25640#
89	77		Ir	−13170#
101	66	167	Dy	−59940
100	67		Ho	−62292
99	68		Er	−63299
98	69		Tm	−62552
97	70		Yb	−60598
96	71		Lu	−57470
95	72		Hf	−53470#
94	73		Ta	−48370#
93	74		W	−42370#
92	75		Re	−34850#
91	76		Os	−26740#
90	77		Ir	−17140#
101	67	168	Ho	−60280
100	68		Er	−62999
99	69		Tm	−61321
98	70		Yb	−61578
97	71		Lu	−57110
96	72		Hf	−55210#
95	73		Ta	−48610#
94	74		W	−44910#
93	75		Re	−35820#
92	76		Os	−30110
91	77		Ir	−18560#
90	78		Pt	−11150#
102	67	169	Ho	−58806
101	68		Er	−60931
100	69		Tm	−61282
99	70		Yb	−60374
98	71		Lu	−58081
97	72		Hf	−54730
96	73		Ta	−50280#
95	74		W	−45020#
94	75		Re	−38450#
93	76		Os	−30880#
92	77		Ir	−22140#
91	78		Pt	−12600#
103	67	170	Ho	−56250
102	68		Er	−60118
101	69		Tm	−59804
100	70		Yb	−60772
99	71		Lu	−57332
98	72		Hf	−56130#
97	73		Ta	−50330#
96	74		W	−47240#
95	75		Re	−39090#
94	76		Os	−33934
93	77		Ir	−23320#
92	78		Pt	−16510#
103	68	171	Er	−57728
102	69		Tm	−59219
101	70		Yb	−59315
100	71		Lu	−57836
99	72		Hf	−55440#
98	73		Ta	−51540#
97	74		W	−47240#
96	75		Re	−41340#
95	76		Os	−34570#
94	77		Ir	−26360#
93	78		Pt	−17710#
104	68	172	Er	−56493
103	69		Tm	−57383
102	70		Yb	−59264
101	71		Lu	−56743
100	72		Hf	−56390
99	73		Ta	−51470
98	74		W	−48970#
97	75		Re	−41680#
96	76		Os	−37260#
95	77		Ir	−27430#
94	78		Pt	−21220

N	Z	A	EL	Mass excess (keV)
105	68	173	Er	−53770
104	69		Tm	−56267
103	70		Yb	−57560
102	71		Lu	−56887
101	72		Hf	−55290#
100	73		Ta	−52490#
99	74		W	−48710#
98	75		Re	−43560#
97	76		Os	−37540#
96	77		Ir	−30220#
95	78		Pt	−22100#
94	79		Au	−12840#
105	69	174	Tm	−53860
104	70		Yb	−56953
103	71		Lu	−55577
102	72		Hf	−55849
101	73		Ta	−51850#
100	74		W	−50150#
99	75		Re	−43610#
98	76		Os	−39950#
97	77		Ir	−31060#
96	78		Pt	−25326
95	79		Au	−14210#
106	69	175	Tm	−52300
105	70		Yb	−54704
104	71		Lu	−55173
103	72		Hf	−54486
102	73		Ta	−52490#
101	74		W	−49590#
100	75		Re	−45240#
99	76		Os	−40070#
98	77		Ir	−33400#
97	78		Pt	−25960#
96	79		Au	−17340#
95	80		Hg	−8270#
107	69	176	Tm	−49600#
106	70		Yb	−53502
105	71		Lu	−53394.6
104	72		Hf	−54581
103	73		Ta	−51480
102	74		W	−50680#
101	75		Re	−45180#
100	76		Os	−42030#
99	77		Ir	−34020#
98	78		Pt	−28940#
97	79		Au	−18570#
96	80		Hg	−11890
107	70	177	Yb	−50997
106	71		Lu	−52395.4
105	72		Hf	−52893.2
104	73		Ta	−51735
103	74		W	−49730#
102	75		Re	−46230#
101	76		Os	−41930#
100	77		Ir	−36000#
99	78		Pt	−29470#
98	79		Au	−21500#
97	80		Hg	−12940#
108	70	178	Yb	−49706
107	71		Lu	−50336
106	72		Hf	−52447.4
105	73		Ta	−50540
104	74		W	−50450
103	75		Re	−45790
102	76		Os	−43550#
101	77		Ir	−36290#
100	78		Pt	−31950#
99	79		Au	−22580#
98	80		Hg	−16323
108	71	179	Lu	−49130
107	72		Hf	−50476.0
106	73		Ta	−50366
105	74		W	−49307
104	75		Re	−46620
103	76		Os	−43010#
102	77		Ir	−38020#
101	78		Pt	−32350#
100	79		Au	−24990#
99	80		Hg	−17110#
98	81		Tl	−7920#
109	71	180	Lu	−46690
108	72		Hf	−49792.8
107	73		Ta	−48940
106	74		W	−49648
105	75		Re	−45850
104	76		Os	−44350#
103	77		Ir	−37950#
102	78		Pt	−34350#
101	79		Au	−25800#
100	80		Hg	−20260#
99	81		Tl	−9350#
109	72	181	Hf	−47417.1
108	73		Ta	−48445
107	74		W	−48259
106	75		Re	−46560#
105	76		Os	−43530#
104	77		Ir	−39460#
103	78		Pt	−34380#
102	79		Au	−27830#
101	80		Hg	−20760#
100	81		Tl	−12320#
110	72	182	Hf	−46063
109	73		Ta	−46437
108	74		W	−48250
107	75		Re	−45450
106	76		Os	−44600
105	77		Ir	−39150#
104	78		Pt	−36180#
103	79		Au	−28330#
102	80		Hg	−23520#
101	81		Tl	−13550#
111	72	183	Hf	−43290
110	73		Ta	−45299
109	74		W	−46370
108	75		Re	−45814
107	76		Os	−43510#
106	77		Ir	−40320#
105	78		Pt	−35740#
104	79		Au	−30130#
103	80		Hg	−23890#
102	81		Tl	−16170#
101	82		Pb	−7730#
112	72	184	Hf	−41500
111	73		Ta	−42844
110	74		W	−45710
109	75		Re	−44218
108	76		Os	−44257
107	77		Ir	−39540
106	78		Pt	−37330#
105	79		Au	−30240#
104	80		Hg	−26260#
103	81		Tl	−17070#
102	82		Pb	−11060#
112	73	185	Ta	−41403
111	74		W	−43393
110	75		Re	−43826
109	76		Os	−42811
108	77		Ir	−40310#
107	78		Pt	−36610#
106	79		Au	−31850#
105	80		Hg	−26170#
104	81		Tl	−19400#
103	82		Pb	−11650#
113	73	186	Ta	−38620
112	74		W	−42517
111	75		Re	−41933
110	76		Os	−43007
109	77		Ir	−39176
108	78		Pt	−37850
107	79		Au	−31580#
106	80		Hg	−28550#
105	81		Tl	−20020#
104	82		Pb	−14630#
113	74	187	W	−39912
112	75		Re	−41224
111	76		Os	−41227
110	77		Ir	−39730#
109	78		Pt	−36830#
108	79		Au	−33110#
107	80		Hg	−28170#
106	81		Tl	−22130#
105	82		Pb	−15060#

N	Z	A	EL	Mass excess (keV)
114	74	188	W	−38676
113	75		Re	−39025
112	76		Os	−41145
111	77		Ir	−38350
110	78		Pt	−37832
109	79		Au	−32530#
108	80		Hg	−30200#
107	81		Tl	−22470#
106	82		Pb	−17720#
105	83		Bi	−7380#
115	74	189	W	−35490
114	75		Re	−37987
113	76		Os	−38995
112	77		Ir	−38460
111	78		Pt	−36499
110	79		Au	−33800#
109	80		Hg	−29600#
108	81		Tl	−24400#
107	82		Pb	−17900#
106	83		Bi	−9710#
116	74	190	W	−34270
115	75		Re	−35540
114	76		Os	−38717
113	77		Ir	−36720
112	78		Pt	−37338
111	79		Au	−32896
110	80		Hg	−31300#
109	81		Tl	−24700#
108	82		Pb	−20430#
107	83		Bi	−10730#
116	75	191	Re	−34361
115	76		Os	−36403
114	77		Ir	−36716
113	78		Pt	−35710
112	79		Au	−33880
111	80		Hg	−30540
110	81		Tl	−26240#
109	82		Pb	−20340#
108	83		Bi	−12940#
117	75	192	Re	−31790#
116	76		Os	−35893
115	77		Ir	−34857
114	78		Pt	−36311
113	79		Au	−32796
112	80		Hg	−32000#
111	81		Tl	−25970#
110	82		Pb	−22550#
109	83		Bi	−13600#
108	84		Po	−7980#
117	76	193	Os	−33406
116	77		Ir	−34543
115	78		Pt	−34487
114	79		Au	−33490#
113	80		Hg	−31150#
112	81		Tl	−27460
111	82		Pb	−22250#
110	83		Bi	−15660#
109	84		Po	−8370#
118	76	194	Os	−32441
117	77		Ir	−32538
116	78		Pt	−34787
115	79		Au	−32278
114	80		Hg	−32238
113	81		Tl	−27090#
112	82		Pb	−24240#
111	83		Bi	−16260#
110	84		Po	−11010#
109	85		At	−810#
119	76	195	Os	−29700
118	77		Ir	−31702
117	78		Pt	−32821
116	79		Au	−32591
115	80		Hg	−31070
114	81		Tl	−28290
113	82		Pb	−23770#
112	83		Bi	−17980#
111	84		Po	−11170#
110	85		At	−3110#
120	76	196	Os	−28300
119	77		Ir	−29460
118	78		Pt	−32671
117	79		Au	−31165
116	80		Hg	−31851
115	81		Tl	−27520#
114	82		Pb	−25450#
113	83		Bi	−17990
112	84		Po	−13470#
111	85		At	−3970#
120	77	197	Ir	−28290
119	78		Pt	−30446
118	79		Au	−31165
117	80		Hg	−30565
116	81		Tl	−28420
115	82		Pb	−24820#
114	83		Bi	−19670
113	84		Po	−13410#
112	85		At	−6140#
121	77	198	Ir	−25930#
120	78		Pt	−29930
119	79		Au	−29606
118	80		Hg	−30979
117	81		Tl	−27520
116	82		Pb	−26120#
115	83		Bi	−19570
114	84		Po	−15500#
113	85		At	−6940#
121	78	199	Pt	−27430
120	79		Au	−29119
119	80		Hg	−29571
118	81		Tl	−28070
117	82		Pb	−25270
116	83		Bi	−20940
115	84		Po	−15270#
114	85		At	−8770#
113	86		Rn	−1610#
122	78	200	Pt	−26625
121	79		Au	−27320
120	80		Hg	−29529
119	81		Tl	−27075
118	82		Pb	−26270#
117	83		Bi	−20410
116	84		Po	−17040#
115	85		At	−8970
114	86		Rn	−4000#
123	78	201	Pt	−23750
122	79		Au	−26413
121	80		Hg	−27687
120	81		Tl	−27205
119	82		Pb	−25300
118	83		Bi	−21490
117	84		Po	−16590#
116	85		At	−10770
115	86		Rn	−4120#
114	87		Fr	3830#
123	79	202	Au	−24370#
122	80		Hg	−27370
121	81		Tl	−26003
120	82		Pb	−25957
119	83		Bi	−20810
118	84		Po	−17990#
117	85		At	−10790
116	86		Rn	−6310#
115	87		Fr	3080#
124	79	203	Au	−23153
123	80		Hg	−25292
122	81		Tl	−25784
121	82		Pb	−24811
120	83		Bi	−21590
119	84		Po	−17350
118	85		At	−12310
117	86		Rn	−6220#
116	87		Fr	930#
125	79	204	Au	−20220
124	80		Hg	−24716
123	81		Tl	−24369
122	82		Pb	−25132
121	83		Bi	−20740
120	84		Po	−18360#
119	85		At	−11920
118	86		Rn	−8070#
117	87		Fr	630
116	88		Ra	6020#
125	80	205	Hg	−22312
124	81		Tl	−23846
123	82		Pb	−23792
122	83		Bi	−21085
121	84		Po	−17560
120	85		At	−13050
119	86		Rn	−7780#
118	87		Fr	−1290
117	88		Ra	5800#

N	Z	A	EL	Mass excess (keV)
126	80	206	Hg	−20969
125	81		Tl	−22278
124	82		Pb	−23809
123	83		Bi	−20048
122	84		Po	−18206
121	85		At	−12500
120	86		Rn	−9180#
119	87		Fr	−1440
118	88		Ra	3540#
127	80	207	Hg	−16270
126	81		Tl	−21048
125	82		Pb	−22476
124	83		Bi	−20078
123	84		Po	−17168
122	85		At	−13290
121	86		Rn	−8670
120	87		Fr	−2980
119	88		Ra	3480#
127	81	208	Tl	−16778
126	82		Pb	−21772
125	83		Bi	−18894
124	84		Po	−17492
123	85		At	−12560
122	86		Rn	−9680#
121	87		Fr	−2720
120	88		Ra	1630#
128	81	209	Tl	−13662
127	82		Pb	−17638
126	83		Bi	−18282
125	84		Po	−16391
124	85		At	−12902
123	86		Rn	−8970
122	87		Fr	−3840
121	88		Ra	1790#
120	89		Ac	8870
129	81	210	Tl	−9263
128	82		Pb	−14752
127	83		Bi	−14815
126	84		Po	−15977
125	85		At	−11992
124	86		Rn	−9623
123	87		Fr	−3410
122	88		Ra	400#
121	89		Ac	8590
129	82	211	Pb	−10493
128	83		Bi	−11872
127	84		Po	−12457
126	85		At	−11672
125	86		Rn	−8779
124	87		Fr	−4200
123	88		Ra	800
122	89		Ac	7070

N	Z	A	EL	Mass excess (keV)
130	82	212	Pb	−7572
129	83		Bi	−8146
128	84		Pc	−10394
127	85		At	−8640
126	86		Rn	−8682
125	87		Fr	−3610
124	88		Ra	−220#
123	89		Ac	7230
122	90		Th	12000#
131	82	213	Pb	−3250#
130	83		Bi	−5254
129	84		Po	−6676
128	85		At	−6603
127	86		Rn	−5723
126	87		Fr	−3573
125	88		Ra	310
124	89		Ac	6090
123	90		Th	12060#
132	82	214	Pb	−187.9
131	83		Bi	−1219
130	84		Po	−4494
129	85		At	−3403
128	86		Rn	−4342
127	87		Fr	−980
126	88		Ra	74
125	89		Ac	6370
124	90		Th	10650#
132	83	215	Bi	1710
131	84		Po	−542
130	85		At	−1269
129	86		Rn	−1192
128	87		Fr	289
127	88		Ra	2510
126	89		Ac	5970
125	90		Th	10890
124	91		Pa	17660
133	83	216	Bi	5960#
132	84		Po	1759
131	85		At	2226
130	86		Rn	232
129	87		Fr	2960
128	88		Ra	3269
127	89		Ac	8060
126	90		Th	10270#
125	91		Pa	17660
133	84	217	Po	5830#
132	85		At	4373
131	86		Rn	3634
130	87		Fr	4293
129	88		Ra	5863
128	89		Ac	8684
127	90		Th	12160
126	91		Pa	17000

N	Z	A	EL	Mass excess (keV)
134	84	218	Po	8351.7
133	85		At	8089
132	86		Rn	5198
131	87		Fr	7036
130	88		Ra	6630
129	89		Ac	10820
128	90		Th	12346
127	91		Pa	18590
134	85	219	At	10520
133	86		Rn	8829
132	87		Fr	8609
131	88		Ra	9365
130	89		Ac	11540
129	90		Th	14450
128	91		Pa	18490#
135	85	220	At	14290#
134	86		Rn	10589
133	87		Fr	11451
132	88		Ra	10250
131	89		Ac	13730
130	90		Th	14646
129	91		Pa	20180#
135	86	221	Rn	14410#
134	87		Fr	13255
133	88		Ra	12938
132	89		Ac	14500
131	90		Th	16916
130	91		Pa	20310#
136	86	222	Rn	16367.0
135	87		Fr	16360
134	88		Ra	14301
133	89		Ac	16603
132	90		Th	17183
131	91		Pa	21940
136	87	223	Fr	18381
135	88		Ra	17234
134	89		Ac	17818
133	90		Th	19243
132	91		Pa	22310
137	87	224	Fr	21630
136	88		Ra	18803
135	89		Ac	20200
134	90		Th	19980
133	91		Pa	23780
138	87	225	Fr	23840
137	88		Ra	21987
136	89		Ac	21615
135	90		Th	22283
134	91		Pa	24310

N	Z	A	EL	Mass excess (keV)
139	87	226	Fr	27200
138	88		Ra	23662.6
137	89		Ac	24298
136	90		Th	23180
135	91		Pa	26015
134	92		U	27170
140	87	227	Fr	29590
139	88		Ra	27172.6
138	89		Ac	25849.0
137	90		Th	25805
136	91		Pa	26825
135	92		U	28870#
141	87	228	Fr	33140#
140	88		Ra	28936
139	89		Ac	28890
138	90		Th	26748
137	91		Pa	28852
136	92		U	29208
141	88	229	Ra	32480
140	89		Ac	30720
139	90		Th	29580
138	91		Pa	29876
137	92		U	31181
136	93		Np	33740
142	88	230	Ra	34460#
141	89		Ac	33760#
140	90		Th	30858.6
139	91		Pa	32162
138	92		U	31598
137	93		Np	35220
142	89	231	Ac	35910
141	90		Th	33812.1
140	91		Pa	33422.2
139	92		U	33780
138	93		Np	35620
137	94		Pu	38390#
143	89	232	Ac	39240#
142	90		Th	35444.4
141	91		Pa	35923
140	92		U	34586
139	93		Np	37280#
138	94		Pu	38349
143	90	233	Th	38729.4
142	91		Pa	37485.9
141	92		U	36914
140	93		Np	38000#
139	94		Pu	40020
138	95		Am	43170#
144	90	234	Th	40607
143	91		Pa	40337
142	92		U	38142.0
141	93		Np	39950
140	94		Pu	40333
139	95		Am	44340#

N	Z	A	EL	Mass excess (keV)
145	90	235	Th	44250
144	91		Pa	42320
143	92		U	40915.5
142	93		Np	41038.7
141	94		Pu	42160
140	95		Am	44640#
139	96		Cm	48020#
145	91	236	Pa	45540
144	92		U	42441.7
143	93		Np	43370
142	94		Pu	42879
141	95		Am	46000#
140	96		Cm	47870#
146	91	237	Pa	47640
145	92		U	45387.2
144	93		Np	44868.3
143	94		Pu	45086
142	95		Am	46630#
141	96		Cm	49150#
140	97		Bk	53190#
147	91	238	Pa	51270
146	92		U	47306.0
145	93		Np	47451.6
144	94		Pu	46160.2
143	95		Am	48420
142	96		Cm	49390
141	97		Bk	54170#
147	92	239	U	50570.9
146	93		Np	49306.9
145	94		Pu	48584.9
144	95		Am	49385
143	96		Cm	51090#
142	97		Bk	54270#
141	98		Cf	58240#
148	92	240	U	52711
147	93		Np	52210
146	94		Pu	50122.4
145	95		Am	51491
144	96		Cm	51701
143	97		Bk	55600#
142	98		Cf	58020#
148	93	241	Np	54260
147	94		Pu	52952.0
146	95		Am	52931.2
145	96		Cm	53696
144	97		Bk	56100#
143	98		Cf	59170#
142	99		Es	63820#
149	93	242	Np	57410
148	94		Pu	54713.9
147	95		Am	55463.2
146	96		Cm	54800.7
145	97		Bk	57700#
144	98		Cf	59330
143	99		Es	64690#

N	Z	A	EL	Mass excess (keV)
150	93	243	Np	59921
149	94		Pu	57751
148	95		Am	57171
147	96		Cm	57177.3
146	97		Bk	58682
145	98		Cf	60910#
144	99		Es	64720#
143	100		Fm	69360#
150	94	244	Pu	59801
149	95		Am	59876.3
148	96		Cm	58449.0
147	97		Bk	60690
146	98		Cf	61459
145	99		Es	65960#
144	100		Fm	69040#
151	94	245	Pu	63174
150	95		Am	61893
149	96		Cm	60998.0
148	97		Bk	61812
147	98		Cf	63377
146	99		Es	66380#
145	100		Fm	70100#
152	94	246	Pu	65365
151	95		Am	64991
150	96		Cm	62613.3
149	97		Bk	64010#
148	98		Cf	64087.3
147	99		Es	67930#
146	100		Fm	70130
152	95	247	Am	67230#
151	96		Cm	65528
150	97		Bk	65485
149	98		Cf	66150#
148	99		Es	68550
147	100		Fm	71540#
146	101		Md	76040#
153	95	248	Am	70590#
152	96		Cm	67388
151	97		Bk	68099
150	98		Cf	67239
149	99		Es	70270#
148	100		Fm	71885
147	101		Md	77080#
153	96	249	Cm	70746
152	97		Bk	69843.6
151	98		Cf	69717.9
150	99		Es	71110
149	100		Fm	73500#
148	101		Md	77260#
154	96	250	Cm	72985
153	97		Bk	72948
152	98		Cf	71167.0
151	99		Es	73270#
150	100		Fm	74063
149	101		Md	78600#

N	Z	A	EL	Mass excess (keV)
155	96	251	Cm	76650#
154	97		Bk	75230#
153	98		Cf	74128
152	99		Es	74507
151	100		Fm	76000#
150	101		Md	79020#
149	102		No	82780#
155	97	252	Bk	78530#
154	98		Cf	76030
153	99		Es	77263
152	100		Fm	76817
151	101		Md	80540#
150	102		No	82856
155	98	253	Cf	79296
154	99		Es	79007.7
153	100		Fm	79339
152	101		Md	81240#
151	102		No	84330#
150	103		Lr	88640#
156	98	254	Cf	81337
155	99		Es	81990
154	100		Fm	80897
153	101		Md	83490#
152	102		No	84723
151	103		Lr	89730#
156	99	255	Es	84090#
155	100		Fm	83787
154	101		Md	84842
153	102		No	86870#
152	103		Lr	90050#
151	104		Rf	94310#
157	99	256	Es	87160#
156	100		Fm	85482
155	101		Md	87522
154	102		No	87796
153	103		Lr	91740#
152	104		Rf	94280#
157	100	257	Fm	88585
156	101		Md	89030#
155	102		No	90220
154	103		Lr	92670#
153	104		Rf	95890#
152	105		Ha	100390#
157	101	258	Md	91820#
156	102		No	91420#
155	103		Lr	94750#
154	104		Rf	96350#
153	105		Ha	101550#
157	102	259	No	94018
156	103		Lr	95850
155	104		Rf	98300#
154	105		Ha	102070#

N	Z	A	EL	Mass excess (keV)
157	103	260	Lr	98100
156	104		Rf	99020#
155	105		Ha	103440#
154	106		Nh	106910#
157	104	261	Rf	101240#
156	105		Ha	104160#
155	106		Nh	108220#
157	105	262	Ha	105970#
156	106		Nh	108470#
155	107		Ns	114510#
157	106	263	Nh	110120#
156	107		Ns	114800#
157	107	264	Ns	115960#
156	108		Uo	120130#
157	108	265	Uo	121240#
157	109	266	Ue	128210#

INDEX